IET POWER AND ENERGY SERIES 73

Wide Area Monitoring, Protection and Control Systems

Other volumes in this series:

Volume 1 **Power Circuit Breaker Theory and Design** C.H. Flurscheim (Editor)
Volume 4 **Industrial Microwave Heating** A.C. Metaxas and R.J. Meredith
Volume 7 **Insulators for High Voltages** J.S.T. Looms
Volume 8 **Variable Frequency AC Motor Drive Systems** D. Finney
Volume 10 **SF_6 Switchgear** H.M. Ryan and G.R. Jones
Volume 11 **Conduction and Induction Heating** E.J. Davies
Volume 13 **Statistical Techniques for High Voltage Engineering** W. Hauschild and W. Mosch
Volume 14 **Uninterruptible Power Supplies** J. Platts and J.D. St Aubyn (Editors)
Volume 15 **Digital Protection for Power Systems** A.T. Johns and S.K. Salman
Volume 16 **Electricity Economics and Planning** T.W. Berrie
Volume 18 **Vacuum Switchgear** A. Greenwood
Volume 19 **Electrical Safety: a guide to causes and prevention of hazards** J. Maxwell Adams
Volume 21 **Electricity Distribution Network Design, 2nd Edition** E. Lakervi and E.J. Holmes
Volume 22 **Artificial Intelligence Techniques in Power Systems** K. Warwick, A.O. Ekwue and R. Aggarwal (Editors)
Volume 24 **Power System Commissioning and Maintenance Practice** K. Harker
Volume 25 **Engineers' Handbook of Industrial Microwave Heating** R.J. Meredith
Volume 26 **Small Electric Motors** H. Moczala *et al.*
Volume 27 **AC-DC Power System Analysis** J. Arrillaga and B.C. Smith
Volume 29 **High Voltage Direct Current Transmission, 2nd Edition** J. Arrillaga
Volume 30 **Flexible AC Transmission Systems (FACTS)** Y-H. Song (Editor)
Volume 31 **Embedded generation** N. Jenkins *et al.*
Volume 32 **High Voltage Engineering and Testing, 2nd Edition** H.M. Ryan (Editor)
Volume 33 **Overvoltage Protection of Low-Voltage Systems, Revised Edition** P. Hasse
Volume 36 **Voltage Quality in Electrical Power Systems** J. Schlabbach *et al.*
Volume 37 **Electrical Steels for Rotating Machines** P. Beckley
Volume 38 **The Electric Car: Development and future of battery, hybrid and fuel-cell cars** M. Westbrook
Volume 39 **Power Systems Electromagnetic Transients Simulation** J. Arrillaga and N. Watson
Volume 40 **Advances in High Voltage Engineering** M. Haddad and D. Warne
Volume 41 **Electrical Operation of Electrostatic Precipitators** K. Parker
Volume 43 **Thermal Power Plant Simulation and Control** D. Flynn
Volume 44 **Economic Evaluation of Projects in the Electricity Supply Industry** H. Khatib
Volume 45 **Propulsion Systems for Hybrid Vehicles** J. Miller
Volume 46 **Distribution Switchgear** S. Stewart
Volume 47 **Protection of Electricity Distribution Networks, 2nd Edition** J. Gers and E. Holmes
Volume 48 **Wood Pole Overhead Lines** B. Wareing
Volume 49 **Electric Fuses, 3rd Edition** A. Wright and G. Newbery
Volume 50 **Wind Power Integration: Connection and system operational aspects** B. Fox *et al.*
Volume 51 **Short Circuit Currents** J. Schlabbach
Volume 52 **Nuclear Power** J. Wood
Volume 53 **Condition Assessment of High Voltage Insulation in Power System Equipment** R.E. James and Q. Su
Volume 55 **Local Energy: Distributed generation of heat and power** J. Wood
Volume 56 **Condition Monitoring of Rotating Electrical Machines** P. Tavner, L. Ran, J. Penman and H. Sedding
Volume 57 **The Control Techniques Drives and Controls Handbook, 2nd Edition** B. Drury
Volume 58 **Lightning Protection** V. Cooray (Editor)
Volume 59 **Ultracapacitor Applications** J.M. Miller
Volume 62 **Lightning Electromagnetics** V. Cooray
Volume 63 **Energy Storage for Power Systems, 2nd Edition** A. Ter-Gazarian
Volume 65 **Protection of Electricity Distribution Networks, 3rd Edition** J. Gers
Volume 66 **High Voltage Engineering Testing, 3rd Edition** H. Ryan (Editor)
Volume 67 **Multicore Simulation of Power System Transients** F.M. Uriate
Volume 68 **Distribution System Analysis and Automation** J. Gers
Volume 69 **The Lightening Flash, 2nd Edition** V. Cooray (Editor)
Volume 70 **Economic Evaluation of Projects in the Electricity Supply Industry, 3rd Edition** H. Khatib
Volume 72 **Control Circuits in Power Electronics: Practical issues in design and implementation** M. Castilla (Editor)
Volume 74 **Power Electronic Converters and Systems: Frontiers and applications** A. M. Trzynadlowski (Editor)
Volume 75 **Power Distribution Automation** B. Das (Editor)
Volume 76 **Power System Stability: Modelling, analysis and control** B. Om P. Malik
Volume 78 **Numerical Analysis of Power System Transients and Dynamics** A. Ametani (Editor)
Volume 79 **Vehicle-to-Grid: Linking electric vehicles to the smart grid** J. Lu and J. Hossain (Editors)
Volume 86 **Advances in Power System Modelling, Control and Stability Analysis** F. Milano (Editor)
Volume 905 **Power System Protection, 4 volumes**

Wide Area Monitoring, Protection and Control Systems

The enabler for Smarter Grids

Edited by
Alfredo Vaccaro and
Ahmed Faheem Zobaa

The Institution of Engineering and Technology

Published by The Institution of Engineering and Technology, London, United Kingdom

The Institution of Engineering and Technology is registered as a Charity in England & Wales (no. 211014) and Scotland (no. SC038698).

First published 2016

The Institution of Engineering and Technology
Michael Faraday House
Six Hills Way, Stevenage
Herts, SG1 2AY, United Kingdom

www.theiet.org

British Library Cataloguing in Publication Data
A catalogue record for this product is available from the British Library

ISBN 978-1-84919-830-1 (hardback)
ISBN 978-1-84919-831-8 (PDF)

Typeset in India by MPS Limited
Printed in the UK by CPI Group (UK) Ltd, Croydon

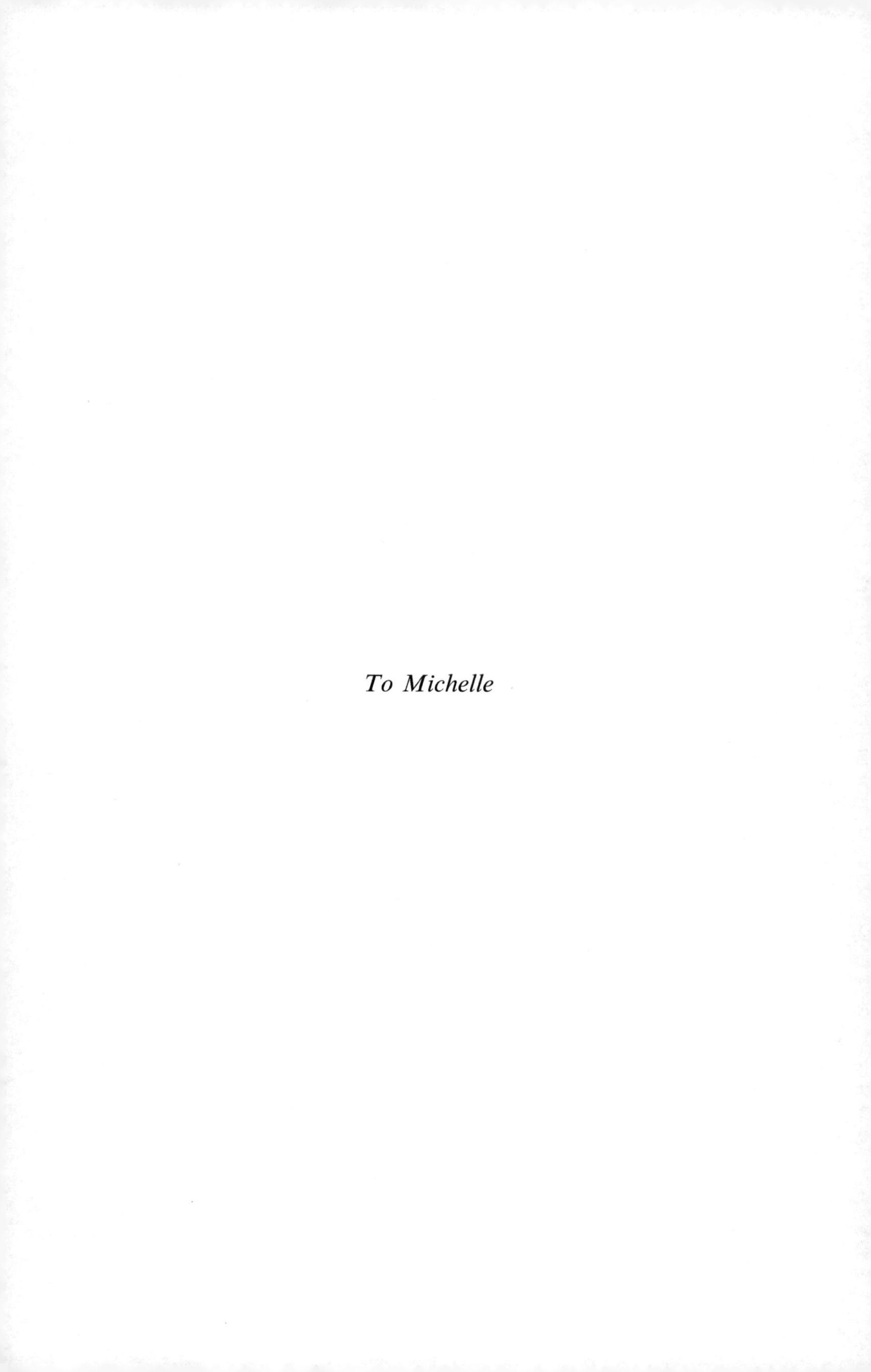

To Michelle

Contents

Preface

Today, electric power transmission systems should face many demanding challenges, which include balancing between reliability, economics, environmental, and other social objectives to optimize the grid assets and satisfy the growing electrical demand.

Moreover, the operational environment of transmission systems is becoming increasingly rigorous due to continually evolving functions of interconnected power networks from operation jurisdiction to control responsibly – coupled with the rising demand and expectation for reliability.

To address these critical issues, wide-area monitoring, protection and control systems (WAMPAC) have been recognized as the most promising enabling technologies. In particular, it is expected that the large-scale deployment of these paradigms could enhance the efficiency of transmission systems and the use of renewable energy resources, by improving, at the same time, the security and reliability levels of interconnected power systems. Moreover, they could support the evolution of transmission networks toward active, flexible, and self-healing web energy networks composed by distributed and cooperative energy resources.

WAMPAC involve the use of system-wide information to avoid large disturbances and reduce the probability of catastrophic events by supporting the application of adaptive protection and control strategies aimed at increasing network capacity and minimizing wide-area disturbances. To this aim WAMPAC requires precise phasor and frequency information, which are acquired by deploying multiple time-synchronized sensors, known as phasor measurement units (PMUs), providing precise synchronized information about voltage and current phasors, frequency and rate-of-change-of-frequency.

WAMPACs can be deployed according to different physical architectures, which are mainly analyzed in Chapter 1. In this chapter, specific focus on the Italian WAMPAC architecture and functionalities is provided in order to demonstrate the role of system-wide information processing in improving the performance of a real and complex power transmission system.

The experimental results discussed in this chapter demonstrate that the WAMPAC performance is strictly influenced by the number and the location of PMUs. Hence, effective mathematical methodologies aimed at finding the minimum number of PMUs for complete system observability represent a relevant issue to address. In solving this challenging issue, the optimal placing methodology should consider the fact that the transmission network topology may change when the power system is affected by a contingency event. Consequently, a robust

WAMPAC, which can ensure complete system observability under the failure of any transmission line or even a PMU, should be resilient to the effect of power system contingencies and loss of measurements, such as the failure of a PMU or its communication links. To solve this problem, Chapter 2 proposes a novel methodology for optimal PMUs placement, which integrates a fault tree analysis for the quantitative reliability evaluation of substation monitoring systems based on branch PMUs. The reliability evaluation is performed considering the PMU and all the additional equipment that are part of the substation monitoring system.

The deployment of an optimal number of PMUs allows WAMPAC to analyze the dynamic behavior of the power system, identify in real-time inter area oscillations, and monitor power flows in interconnected areas. All these features could support the transmission system operators in reliably operating the power transmission system closer to its stability limits, exploiting transmission and generation capacity more efficiently. As a result, renewable power generators can be used more effectively, and the marginal cost of power generation can be reduced.

These benefits have been confirmed in Chapter 3, where the role of wide-area information processing in designing system integrity protection schemes is analyzed. The results presented in this chapter demonstrates as WAMPAC could be considered as an enabling technology to discover and treat real-time disturbances in wide area in order to prevent system blackout and maintain integrity of the whole power system, not just locally or focused on particular element of the grid.

In this domain, the most promising research directions are oriented toward the enhancement of the power system monitoring functions with online tools for power system stability analysis. In traditional power systems, system stability limits are typically computed off-line and stored in databases to be monitored by transmission system operators in the real-time environment. Several sources of uncertainty affect such computations and consequently reasonable stability margins must be taken into account when determining operation limits. Despite these precautions, unplanned outages and planned switching actions may cause operational conditions not considered at operation planning stages, and consequently system operators are left with no pertinent stability information. This complex issue has been addressed in Chapter 4, which analyzes the role of online stability assessment in computing stability limits for large-scale power system based on the actual system condition, decreasing the uncertainty, thus providing more accurate stability operation limits.

As easily understandable, this computing process is a very complex and time-intensive task, since it requires the periodic estimation of the power system state, the analysis of the massive data streams generated by the grid sensors and the repetitive solution of large-scale optimization problems, which are complex, non-linear, and NP-hard problems. Moreover, in order to provide the grid operators with updated information to better understand and reduce the impact of system uncertainties associated with load and generation variations, the required computation times should be fast enough.

As outlined in Chapter 5, this goal can be achieved by means of computing paradigms which can support rapid power systems analysis in the typical context of "big data". The idea is to extract useful information from historical operation data sets, which are expected to grow over and over because of the pervasive deployment of grid sensors. The enabling method is given by a Fuzzy transform-based mathematical kernel, which aims at reducing the cardinality of optimal power flow problems so that more efficient algorithms can be used to get solutions.

An effective deployment of the WAMPAC on real transmission networks requires a comprehensive cyber-security analysis aimed at detecting the potential vulnerabilities of the architecture modules to cyber-attacks, and designing proper reactive process for detecting and mitigating these threats. This issue is addressed in Chapter 6, where the vulnerability of the power system state estimation is investigated through the analysis of the tradeoff between attack magnitude and the attack error. To this aim two false data injection attacks (FDIAs), i.e., perfect and imperfect attacks, are defined and discussed. To detect the FDIAs, a measurement consistency check based index is proposed. This index is defined by the normalized measurement difference between received measurements and the interpolated measurements that are calculated through a small number of selected secure alternative PMU measurements. The results discussed in this chapter demonstrate that the proposed method could be a good candidate to mitigate the effects of cyber security on power system state estimation and wide-area information processing.

A. Vaccaro
A.F. Zobaa

Chapter 1

Wide area measurement system: the enabler for smarter transmission grids

G. Giannuzzi[1] *and C. Pisani*[1]

The mission of a modern transmission system operator (TSO) is to manage a high-voltage power system and guarantee its *safety*, *quality*, and *affordability* over time. To do this, several analyses, properly implemented via specialist software (hereafter applications), have to be executed both on line and off line. These applications require in input more and more system information for predicting power system dynamic evolution, in each operating condition. When the secure and stable system operating is threatened, proper countermeasures have to be properly implemented. Unfortunately this task is not so straightforward due to (*i*) complexity of the involved dynamics which makes difficult to predict the system behavior (*ii*) the technological limits of the traditional supervisory control and data acquisition (SCADA) systems. For these reasons wide area measurement (monitoring) systems (WAMS) and their wide-scale deployment are receiving an increasing attention for some years. WAMS use sophisticated digital recording devices, i.e., phasor measurement units (PMUs), to record and export global positioning system (GPS)-synchronized, high sampling rate (6–60 samples/second) dynamic power system data. A wide area control system based on WAMS is a typical area network control system in which the communication among sensors, actuators, and controllers occurs through a shared band-limited digital communication network [1]. Wide area protection system aims at enhancing the actual interoperability level in a wide area by a proper management of the protection systems [2]. Wide area monitoring protection and control system is the term coined for describing a wide area system which implements monitoring, protection, and control functionalities [3].

Different open problems can be recognized in the power system literature about WAMS. First of all the research of reliable and resilient tools for preparing, observing, and enhancing a comprehensive dynamic system behavior. Furthermore there is a cogent need to perform an integration of neighbor national WAMS, by exchanging measured data excluding market-sensitive data, especially at the European level. The availability of large amounts of measurements coming from the system key points, i.e., from the primary substations, and the availability of

[1]Terna Rete Italia, Roma, IT 00138, Italy

computing power at low cost offers the possibility to perform advanced dynamic security analyses (DSA) in real time.

In this chapter, a general description of the WAMS architectures is given: particular emphasis is done to the main constitutive items. Then a focus on the Italian WAMS architecture and functionalities is provided. The final part of the chapter is devoted to the characterization of the WAMS data and to a discussion about the preprocessing actions to apply on raw data in order to make successful the DSA investigations.

1.1 WAMS: definition and constitutive sub-processes

Wide area monitoring concept has been first introduced by Bonneville Power Administration in the late 1980s [4]. WAMS consists of advanced measurement technology, information tools, and operational infrastructure that facilitate the understanding and management of the increasingly complex behavior exhibited by large power systems [5]. In its present form, a WAMS may be used as a stand-alone infrastructure that complements the grid's conventional SCADA system. As a complementary system, a WAMS is expressly designed to enhance the operator's real-time "situational awareness" that is necessary for safe and reliable grid operation. In a WAMS three different interconnected subprocesses can be identified, namely, data acquisition, data transmitting, and data processing, which are, respectively, performed by measurement systems, communication systems, and energy management systems (EMS) [6].

WAMS acquire power system data from conventional or advanced measurement devices and transmit it through proper communication systems to the control centers where a preprocessing stage aimed at implementing all the functionalities is initially performed. The aforementioned three WAMS subprocesses employ **data resources**, **communication system**, and **applications** as essential items. An elementary WAMS process view is reported in Figure 1.1 [7].

1.1.1 WAMS data resource

Data resources are disseminated along the power systems with a view to collecting very heterogeneous power system data. Although a comprehensive data classification is furnished in Section 1.4, they are first classified into *operational* and *not operational* [8].

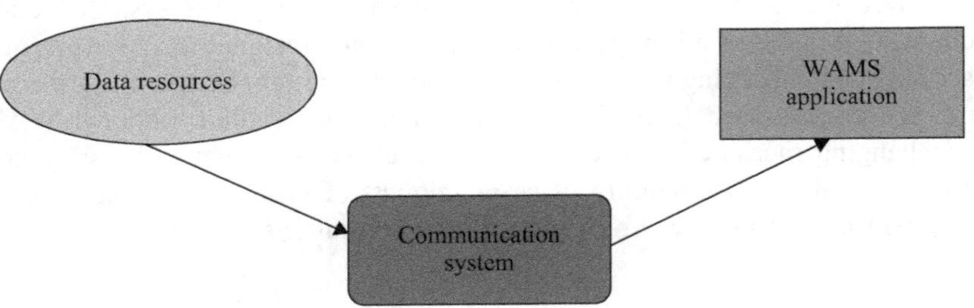

Figure 1.1 WAMS process in power systems

Operational data includes the instantaneous measurements of voltage and current (magnitudes and angles), breaker statuses, etc., recorded by the installed measurement devices and continuously transmitted to the control centers.

Non-operational data includes records or logs of multiple events, e.g., series of faults, power fluctuations, disturbances and lightning strikes recorded by the installed measurement devices and conveyed at a specified time intervals (e.g., multiple of hours) or however at specified conditions.

In a similar manner even data resources in power systems can be classified into operational data resources such as supervisory control and data acquisition, synchronized, phasor measurement system and into non-operational data resources such as circuit breaker monitor, digital fault recorder, and digital protective relay.

1.1.1.1 Supervisory control and data acquisition

SCADA is a computer-assisted system aimed at collecting and processing data by applying operational controls over long distances and so three chief critical functions: data acquisition, supervisory control, and alarm display and control. As far as SCADA hardware is concerned, it includes a master terminal unit (MTU) located in the control centers, one remote field site consisting of either a remote terminal unit (RTU) or programmable logic controllers (PLC) or intelligent electronic devices (IED) and a communication system that provides communication route between remote site and the control center.

MTU is installed in the control center and represents the core of SCADA systems. It performs many functions: (*i*) manages all communications, (*ii*) gathers data of RTU, (*iii*) stores obtained data and information, (*iv*) sends information to other systems, (*v*) commands system actuators that are connected to RTU, and (*vi*) interfaces with operators.

RTU is a stand-alone data acquisition and control unit which monitors and controls equipment at remote sites and transfers data collections to MTU. RTUs are generally based on microprocessor technology and can also be configured as a relay. The RTU size (small, medium, and large) is established on the basis of the analog/digital inputs able to manage.

PLC is a small industrial computer able to perform functions carried out by electrical equipment, e.g., relays, drum switches, and mechanical timer/counters [9, 10]. Since based on a built-in microprocessor technology PLC is more economical, versatile, flexible and configurable, physically compact than the RTU. The communication systems provide communication routes between the master station and the remote sites, through private transmission media (e.g., fiber optic or leased line) or atmospheric means (wireless or satellite).

There are three main physical communication architectures used in SCADA communications: point-to-point, multipoint, and relay station architectures Figure 1.2.

1.1.1.2 Synchronized phasor measurement system

The synchronized phasor measurement system (SPMS) is an advanced device which employing time received from common time source (e.g., GPS) as its sampling clock, is able to measure currents and voltages and calculate the angle

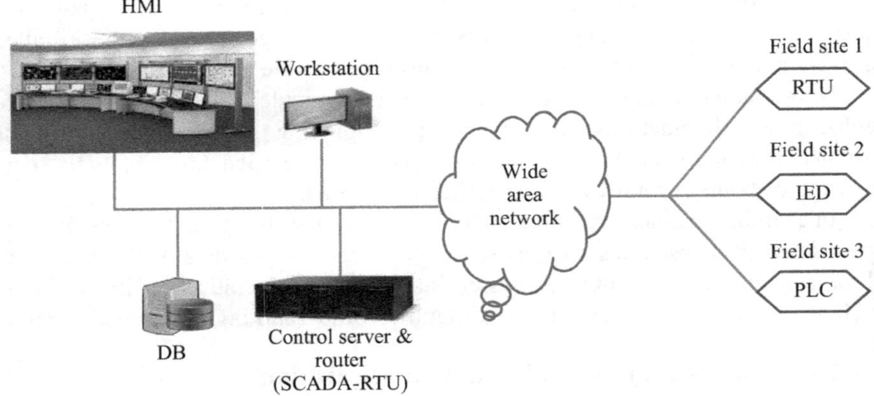

Figure 1.2 Supervisory and control data acquisition architecture

between them. Indeed the set of measured quantities can be customized by adding several further quantities such as local frequency, rates of frequency changes, harmonics measurements, negative and zero sequence quantities [11].

Three main constitutive parts can be identified in a SPMS: phasor measurement unit, phasor data concentrator, and communication system.

PMU is a microprocessor-based device able to acquire the electrical waves (i.e., voltages and currents) on a power system at a typical rate of 48 samples per cycle (2400/2880 samples per second). The resultant time-tagged phasors, also known as *synchrophasors*, can be conveyed to a local or remote receiver at rates up to 60 samples per second (typically 30–60 samples per second).

Basically, a PMU covers in SPMS the same tasks of an RTU in a SCADA system.

Phasor data concentrator (PDC) has a very crucial role in SPMS: It receives synchrophasors from several PMU or others PDC and to fed out them as a single stream by implementing several checks on data consistency. For example, it rejects bad data, aligns the time stamps, and finally creates a coherent record of simultaneously recorded data. At this regard, the main distinguishing element with respect to SCADA architecture relies in the streamed data.

Two major differences can be noted: (*i*) synchrophasor data is continuous and streaming in nature while RTU data is transmitted to the master station either in specified time intervals or when master station requests it; (*ii*) synchrophasor data has a more sensible value than data provided by an RTU which implies stricter requirements for the communication systems (i.e., high bandwidth, low latency communications, etc.), Figure 1.3.

1.1.1.3 Digital fault recorder

Digital fault recorder is a device aimed at archiving highly accurate waveforms regarding fault events. Huge amount of data, related to the three major fault

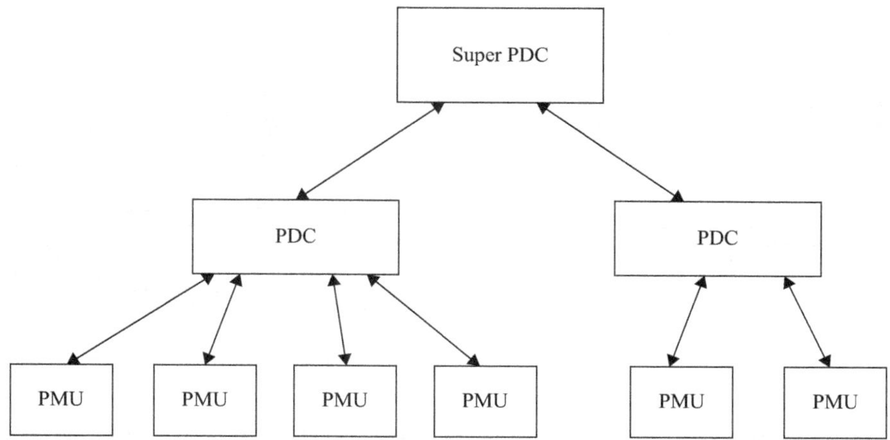

Figure 1.3 Synchronized phasor measurement system structure

dynamic evolution stages, pre-fault, fault, and post-fault stage, are collected such as analog or status data [12]. Typical examples are maximum current, sequence of events, type of fault and the sequence of circuit breakers operations. This data, typically sampled at very high sample rate (about 64 to 356 samples per cycle), cannot be used in real time and hence they are archived as samples for further offline processing.

This task is recently accomplished by using directly SPMS. Actually time-synchronized data provided by PMU, installed for instance on high-voltage power plants bays or critical substations, can allow the fault event reconstruction with sufficient accuracy and precision.

1.1.1.4 Digital protective relay

One of the cogent needs when a fault occurs in a power system is to isolate the area faulty area in order to minimize the impact on the healthy areas. This function is achieved by protective relays which are nowadays of digital type, also known as digital protective relays (DPRs). DPRs are based on advanced microprocessor technology and are able to detect faults in power systems by processing voltage and current waveforms. Even analog and status data communicating with a centralized location can be measured and recorded.

DPR sample rates are quite limited. Nowadays they vary from 64 to 128 samples per cycle [13] due to the compromise need to ensure very fast reactions. Obviously, this implies a lower accuracy of DPR data compared with the other data resources.

1.1.1.5 Circuit breaker monitor

The circuit breaker monitor (CBM) is an electronic device that monitors circuit breakers. The CBM works in real time, capturing detailed information about each

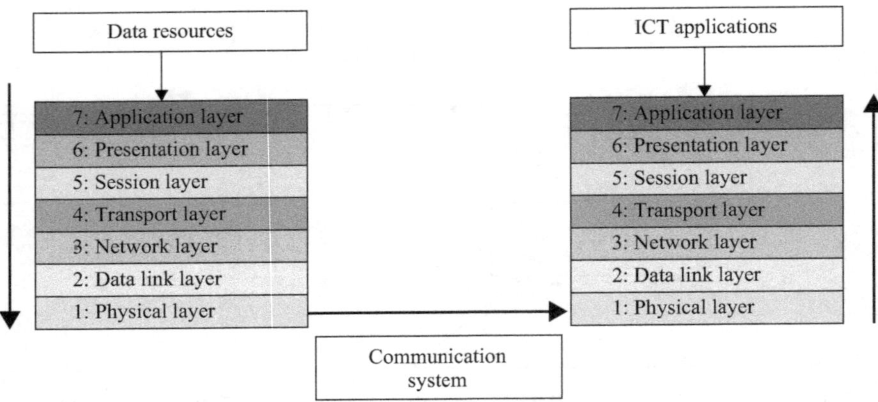

Figure 1.4 Layering in WAMS based on OSI reference model

breaker, for properly acting or manually by the operator or automatically by the protection and control equipment when necessary.

1.1.2 WAMS communication systems

Communication systems play a crucial role in power system operation and control since may put at risk the security of the system enhancing the probability of not supplying customers. Their task in WAMS is to ensure data delivery both from data resources to the control centers and from control centers to the system actuators. An emerging paradigm in new communication systems is open system interconnection (OSI) layer; it deals with a comprehensive architecture for explanation, designing, implementation, standardization and use of communications networks. OSI reference model is formed by seven layers which can be appreciated in Figure 1.4: physical, data link, network, transport, session, presentation, and application layer. WAMS data resources and applications usually interact with the upper layers of the architecture. Actually, Figure 1.4 gives a synthetic vision of the link among OSI layers and data resources, applications and communication system.

WAMS communication systems performance is highly affected by the characteristics of the employed transmission media such as the cost, bandwidth, propagation delay, security, and reliability. Transmission media, as described hereinafter, can be classified as *guided and unguided* [14]. While in guided transmission media the information content travels through a solid medium in an unguided transmission media, the information content is transferred through electromagnetic waves and hence no physical guided transfer path can be identified.

Twisted pair, coaxial cable, power transmission/distribution line, and optical fiber are examples of guided media. Wireless communication such as atmosphere and outer space are examples of unguided media. In the latter case signal strength provided by wireless antenna is more important than the medium itself. Table 1.1

Table 1.1 (a) Wide area monitoring systems guided media

Media	Description	Bandwidth	Latency	Security
Optical fiber	Object linking and embedding for process control (OPC) is employed for its flexibility to be bundled as a cable. In fiber cables, the signal is a light wave either visible or infrared light. Types used in power industries are optical power ground wire and all-dielectric self-supporting.	High	Low	High
Power line carrier	PLC transfers critical communications directly through transmission lines, so a possible failure of the power system infrastructure (e.g., line outage) causes communication problems. According to the data rate values, PLC systems may be classified in two groups: *narrow band* and *broad band.*	Medium	Low	High
Leased line	LL is used together with some technologies for transmitting wide area signals. These technologies are essentially the digital subscriber line, able to provide digital data transmission over leased telephone circuits. According to their data rates and directionality of transmission, to distances to which those rates can be supported and to the size of the wire, several DLS versions can be identified.	Medium	Low–Medium	High

summarizes the most common transmission media used in WAMS by presenting the major characteristics and providing a comparison among them in terms of bandwidth, latency, and security [7, 15].

1.1.3 WAMS applications/functionalities

WAMS applications or functionalities are computer-assisted tools able to process the raw data from data resources presented in Section 1.1.1 in order to extract usable information for system operators in power systems operating and control. Depending on the specific power system interventions, WAMS applications can be classified into *generation*, *transmission*, and *distribution applications*.

Generation applications are aimed at controlling in real time the generator's operation and hence at supervising potential instability phenomena (e.g., transient angle).

Table 1.1 (b) Wide area monitoring systems unguided media

Media	Description	Bandwidth	Latency	Security
WPAN	Wireless personal area network is a network for interconnecting devices located around an individual person's workspace characterized by wireless connection. A WPAN uses some technologies that allow the communication within a very short range (10 m). One of the adopted technologies in WPAN is the bluetooth.	Low–Medium	Low–Medium	Low
WLAN	Wireless local area network connects devices through a wireless distribution method. Wi-Fi is one of the most popular WLAN technologies and it provides high-speed connection on short ranges.	Low–Medium	Medium	Low
WMAN	WiMAX, GPRS, GSM, CDMA, and 3G mobile carrier services are five **wireless metropolitan area network** technologies which are used for WMAN communication. **Worldwide interoperability for microwave access** (WiMAX) is a communication protocol which provides fix and fully mobile data networking. Its theoretical data rate is 70 Mbps with a range of up to a maximum of 50 km with a direct line of sight. Near line of sight conditions seriously limit their range. **Global system for mobile** (GSM) is a standard for mobile telephony system based on circuit-switching technology. Whit this technology connections are "always on". **General packet radio service (GPRS)** is a packet data bearer service over GSM system. It uses a packet radio principle to transfer data at high bandwidth. When a device transmits packages the bandwidth is used, GPRS has hence higher data speed than the GSM. **Code division multi-access** (CDMA) is another data networking technology for mobile communications. It allows all the users to utilize the entire frequency spectrum for all the time. CDMA can create 64 logical	Medium	Medium	Low

Table 1.1 (b) (*Continued*)

Media	Description	Bandwidth	Latency	Security
	channels whereas 8 channels are available in GPRS. **3G mobile carrier services** provide new data carrier services for mobile users. For example, some networks support high speed packet access data communication with HSDPA standard to provide improved downlink speeds.			
WWAN	Wireless wide area network, with satellite communications, may be used in two cases: when a guided medium cannot be established between a remote site and the control center or when there is no line-of-sight between such a remote site and pre-installed communication network. One of the problem of satellites is their high latency that may create serious difficulties for some WAMS applications.	Low–Medium	High	Low

Transmission and subtransmission applications are implemented by group of computer-aided tools also known as EMS. Historically, the main conventional EMS applications were:

- State estimation;
- Load flow;
- Optimal power flow;
- Load forecast;
- Economical dispatch.

Nowadays modern WAMS applications include (and it is not limited to):

- Integrated phasor data platform;
- Wide-area dynamic monitoring and analysis;
- Synchronized disturbance record and replay;
- **Online low-frequency oscillation analysis**;
- Power angle stability prediction and alarming;
- PMU-based state estimation;
- Fault analysis support;
- Comprehensive system load monitoring;
- Dynamic line thermal monitoring;
- Power system restoration support tool.

Therefore, modern EMS are sophisticated both form hardware and software point of view.

Distribution applications, often known as automation applications in *IEEE* community, are "systems that enable a distribution company to monitor, coordinate, and operate distribution components and equipments from remote locations in real time". Their fundamental scope is to reduce costs, improve service availability, and provide better services to the customers. Distribution applications can be classified into substation automation, feeder automation, and consumer-side automation [16].

From a general point of view, WAMS applications can be further classified into two main classes [17]. The first one includes the applications *for local use* meaning tools which use measurements only from their own, area, region, or specific TSO. The second one includes the applications for *wide area use* meaning tools which use measurements from neighboring national WAMS. Furthermore their application execution could be *on line* and *off line*, for instance, for extracting valuable information about power system characteristics on more or less wide time windows (e.g., weakly electromechanical mode damping).

1.2 Italian WAMS architecture and functionalities

The Italian WAMS project started in 2004 with the aim of achieving system benefits in terms of power system security and observability. More specifically, running off line and especially in real time several applications/functionalities, it is able to:

- enhance the TSO awareness about the actual Italian power system so detecting promptly abnormal conditions and avoiding missing alarms;
- provide a comprehensive vision of the Italian power system and neighbors power systems within the European Network of Transmission System Operators for Electricity (ENTSO-e) Continental Synchronous European Area (CESA);
- give a comprehensive observation of highly national and transnational loaded transmission corridors;
- allow the implementation of very fast countermeasures in critical or emergency conditions.

Italian WAMS is devised to host different applications and users. In particular in the central control room of the Terna's National Control Centre (NCC), the real-time trends and alarms related to following applications are delivered:

- Voltage magnitude (low/high);
- Frequency (low/high);
- Angle difference (high);
- Damping (low; from online oscillatory analysis function);
- Voltage collapse;
- Underfrequency load shedding (from load shedding evaluation function);

- Islanding, loss of synchronism, and frequency stability;
- Line thermal loading monitoring;
- SCADA state estimation integration with support of WAM platform.

Terna's NCC is the heart of the Italian power system and is located in the outskirts of Rome. The central control room carries out a detailed control of the country's entire electricity transmission grid. Sophisticated monitoring and management systems allow Terna to control, instant by instant, the electricity produced in Italy or imported from other countries and to safely manage the energy flows. The monitoring activity is carried out non-stop, 24 hours a day, 365 days a year [18].

Figure 1.5 depicts a typical graphical user interface customized on the basis of the Terna experts' suggestions and needs of the operators working in the control room. Italian WAMS is formed by a set of about 100 PMU disseminated along the whole national electricity system, a leased data network, based on direct numerical circuit (DNC) channels and several WAMS applications/functionalities running at NCC. The NCC is able to track in real time the dynamics of the whole ENTSO-e CESA thanks to the data-streaming exchanges with other European TSO. PMU sampling rate is 50 samples per second.

Data is stored in a shared memory having a matrix structure designed to allow short-term data archiving: the rows are associated to a specific time tag (sampling time) and the columns are associated to the different measured quantities by the PMU (such as voltage magnitude and phase, current magnitude and phase, frequency, etc.). The shared memory is sized to allocate a maximum of 512 measures coming from 100 PMU, for a duration of 30 min at a rate of 50 frames-per-second (i.e., one every 20 milliseconds). Such an acquisition here called "fast" is complementary to a "slow" acquisition consisting in a periodic data saving into a relational database. Data management philosophy follows the criterion to free up space from the shared memory by starting from the oldest data which are then available for subsequent off-line evaluations (e.g., inter area oscillation damping evaluation on monthly basis). Therefore the conceived relational database is formed by two levels. The first receives the data from the shared memory every 30 minutes and keeps them for the next 24 hours. The latter keeps the data collected in the last 30 days, with a sampling period of 100 ms and is refreshed on a daily basis.

A comprehensive view of the Italian WAMS architecture can be appreciated by Figure 1.6. From the central system, the real-time acquired data are available to:

- Real-time applications, embedded in the central system;
- Distributed calculations on clients interfaces;
- External calculations on external servers (i.e., Python/MATLAB® platform) via proprietary protocol or standard OPC;
- Visualization clients.

A further specificity of the Italian WAMS architecture relies on a virtual acquisition process which allows feeding the acquisition and storage blocks with

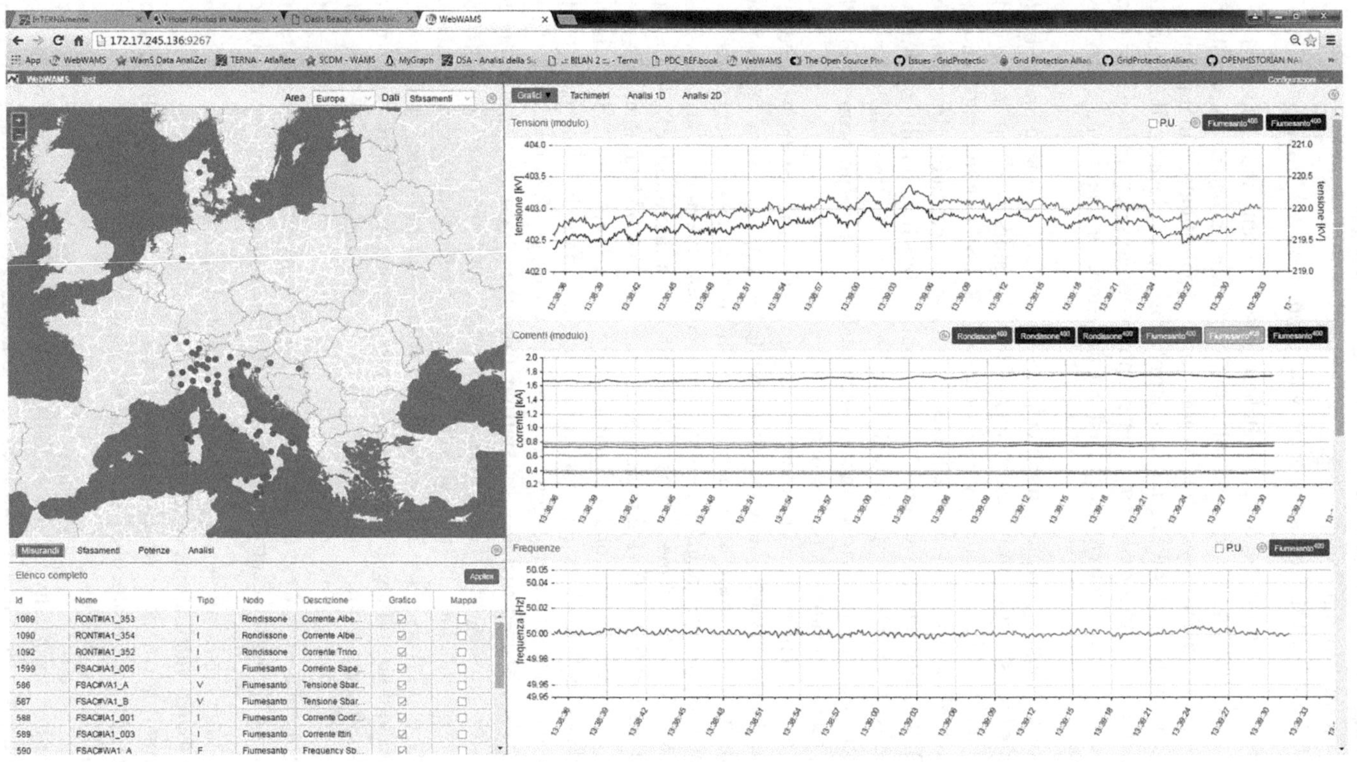

Figure 1.5 Italian WAMS graphical user interface

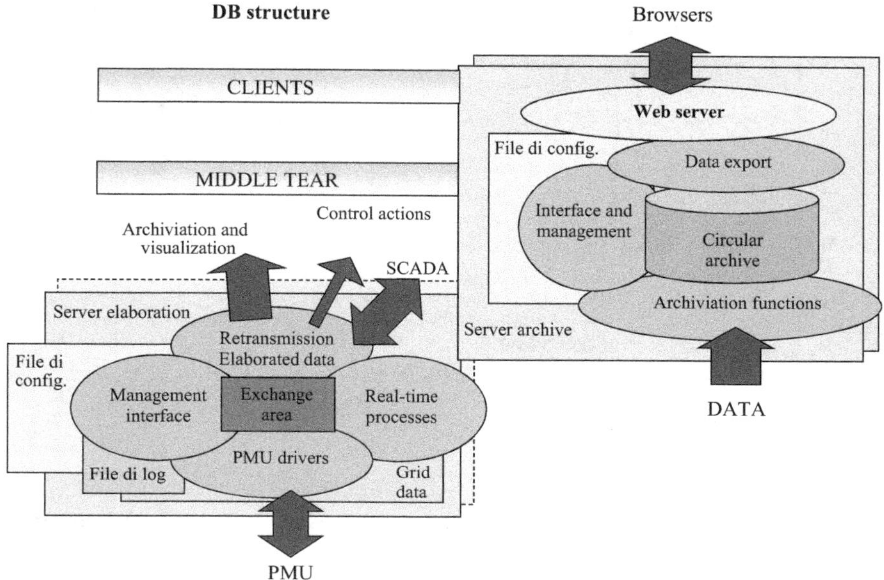

Figure 1.6 Italian WAMS architecture

data coming from sources such as text files. This function is very important since it permits to test the Italian WAMS control system response under critical conditions either actually recorded from the field or specifically produced by simulation. In this manner some operations like alarm functions verification and protections setting can be effectively proven via a noninvasive approach.

1.3 WAMS data classification

Classifying measurement data is particularly important since the performance of the applications discussed in Section 1.2 is dependent on data typology. Each application installed on the EMS of Terna's NCC has been undergone to very long validation phase aimed at assessing its own behavior in different situations which means also with respect to different data typologies. It is quite rare that one algorithm shows good performance with respect to all data typologies for this reason all the ones available in EMS must be selected according to smart selection criteria. From a general point of view, field measurement data can be classified into two categories: *typical* and *nontypical* data [19].

Typical data can be ever be represented by a mathematical model and are responsible of the content information carrying. Nontypical data cannot be represented generally by a meaningful mathematical model (linear) and does not carry information content. Examples of nontypical data are invalid data (NaN) or outliers. Invalid data are very frequent in WAMS registrations since produced, for

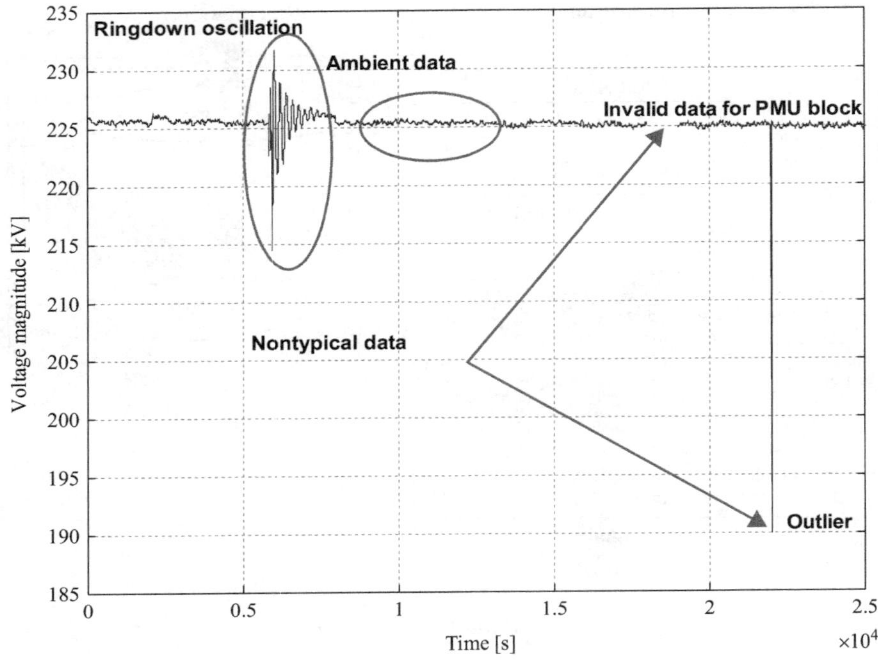

Figure 1.7 WAMS measurement data types

instance, by temporary communication and measurement devices failure. Outliers are instead values quite far from normal waveform trend which may result for instance from a measuring current/voltage transformer misoperation. A special category of nontypical data is represented by transient data right before ringdown oscillations (even in this case a general linear model is not able to describe it).

Although nonexhaustive as per classification, nontypical data can in turn be ranked into *ambient data*, *ringdown oscillation*, and *probing data*.

Ambient data is related to the system working in a neighborhood of the equilibrium point and it is perturbed with a small amplitude random load changes [20]. Ringdown oscillation data is related to the transient occurred after a major disturbance such as a line tripping. It can be easily detected since consisting in an observable oscillation [21]. Probing data is obtained when low-level pseudo-random noise is intentionally injected into the system to test its performance [22].

Figure 1.7 shows an actual voltage magnitude waveform recorded on the 220 kV Italian transmission system. Both typical (ambient and ringdown data) and nontypical data (NaN and outlier) can be clearly appreciated from this figure. The information content carried by the data categories above is different. In order to capture information about the power system status directly from the acquired WAMS measurements, there are some algorithms which work better by using ringdown data while other by using ambient data. Therefore, a smart selection criterion that permits the switching towards those algorithms performing better on this data category at the onset of a transient phenomenon on the power system would be very useful in real-time operation. This is exactly what is devised in Italian WAMS.

1.4 Preprocessing synchronized phasor measurement data for power system analyses

The main traits of the WAMS measurement data have been well described in Section 1.3. Very often the recorded time series are practically unusable as they are for the main applications devised in EMS. In particular, a proper preprocessing stage must be executed preliminarily in order to avoid inconsistent outcomes.

From a general point of view, a preprocessing unit is aimed at [23]:

- Removing defective data;
- Parceling data sets;
- Removing outliers;
- Interpolating missing samples;
- Removing trends;
- Filtering unwanted dynamics.

Missing data has a huge negative impact on the correctness of the applications' outcomes. It is trivial to understand that the greater the magnitude of the data loss, the worst is the algorithm response. The first idea that comes to mind to manage missing data is clearly to exclude them. Indeed this has been rigorously addressed in Reference 24 verifying that if the amount of missing data is limited, indicatively lower than the 6% of the data packet length, to remove missing data and concatenate the correct ones is the more appropriate manner to operate [24]. A different strategy could be to apply an interpolation process to reconstruct the missing parts as proposed in References 25 and 26. On the contrary, if the magnitude of the data loss is not limited, concatenation will introduce artificial transients while interpolation will result in noise, therefore both the actions are to be excluded.

As far as outlier management is concerned, some handling mechanisms have been proposed in References 25, 27–28, which are properly incorporated in the Italian EMS.

Furthermore a trend removal and low-pass filtering, accomplished both with finite impulse response and infinite impulse response filters, is necessary to select the information content (in the frequency domain) strictly necessary for that specific power system analysis.

Figure 1.8 shows the preprocessing chain with the highlighted actions [15].

Defective data removal aims at discarding sequences of identical sample timestamps and/or values, samples with value zero and invalid measurements (e.g., NaN) caused by internal PMU failures occurrence. *Data parceling* aims at removing the missing samples to the data packets to furnish to the filters which are highly sensitive to potential gaps due to missing data. *Removal of outliers* aims at

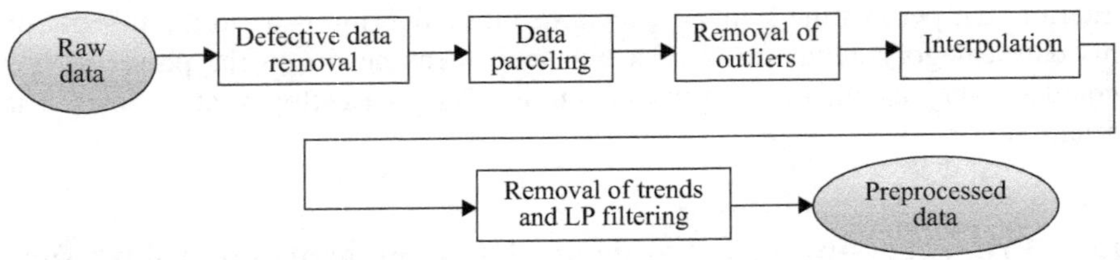

Figure 1.8 Preprocessing unit

avoiding to induce artificial transients when the data is filtered. *Interpolation* aims at replacing missing samples as well as at removing data and outliers so capturing the signal trend: a linear interpolation may be simple and effective. *Removing of mean and low-pass filtering* aims at suppressing the modal content which does not contain important information about the power system small signal stability.

References

[1] P. Hespanha, P. Naghshtabrizi, and X. Yonggang, "A survey of recent results in networked control systems," *Proceedings of the IEEE*, vol. 95, pp. 1, Jan. 2007.

[2] M. M. Begovic, D. Novosel, D. Karlsson, *et al.*, "Wide area protection and emergency control," *IEEE Transaction on Power Delivery*, vol. 93, no. 5, pp. 876–891, May 2005.

[3] V. Terzija, G. Valverde, D. Cay, *et al.*, "Wide area monitoring, protection, and control of future electric power networks", *Proceedings of IEEE*, vol. 99, no. 1, pp. 8093, Jan. 2011.

[4] C.W. Taylor, "Wide area measurement, monitoring and control in power systems," *Workshop on Wide Area Measurement, Monitoring and Control in Power Systems*, Imperial College, London, Mar. 16–17, 2006.

[5] M.D. Hadley, J.B. McBride, T.W. Edgar, L.R. O'Neil, J.D. Johnson, Office of Electricity Delivery and Energy Reliability – Department of Energy – PNLL, "Securing wide area measurement systems", June 2007.

[6] D. Yan, "Wide-area protection and control system with WAMS based", *International Conference on Power System Technology (PowerCon2006)*, China, Oct 22nd-26th, Chongqing, China, 2006, pp. 1–5.

[7] M. Shahraeini and M. H. Javidi, Wide area measurement systems, *Advanced Topics in Measurements*, InTech, available from: http://www.intechopen.com/books/advanced-topics-in-measurements/wide-area-measurement-systems, 2012.

[8] M.S. Thomas, D. Nanda, and I. Ali, "Development of a data warehouse for nonoperational data in power utilities", *Proceedings of Power India Conference*, New Delhi, India, 2006.

[9] G. Clarke and D. Reynders, *Practical Modern SCADA Protocols: DNP3, 60870.5 and Related Systems*. Oxford, England: Elsevier, 2004.

[10] Trends in SCADA for Automated Water Systems, Synchrony, 2001.
[11] A.G. Phadke and J.S. Thorp, *Synchronized Phasor Measurements and Their Applications*, New York, USA: Springer, 2008.
[12] M. Kezunovic, Integration of substation IED information into EMS functionality, final project report, Power Systems Engineering Research Center (PSERC), 2008.
[13] Considerations for use of disturbance recorders, A Report to the System Protection Subcommittee of the Power System Relaying Committee of the IEEE Power Engineering Society, IEEE Inc., 2006.
[14] M. Shahraeini, M.H. Javidi, M.S. Ghazizadeh, "A new approach for classification of data transmission media in power systems", *International Conference on Power System Technology*, October 24–28, Hangzhou, China, 2010, pp. 1–7.
[15] C Pisani, "Real time tracking of electromechanical oscillations in ENTSO-e Continental European Synchronous Area", PhD Dissertation, University of Naples, May 2014.
[16] N. Zhou, Z. Huang, F. Tuffner, *et al.*, *Algorithm summary and evaluation: Automatic implementation of ringdown analysis for electromechanical mode identification from phasor measurements*. Pacific National Northwest Laboratory (PNNL) Report, 2010.
[17] W. Sattinger and G. Giannuzzi, "Monitoring continental Europe: An overview of WAM systems used in Italy and Switzerland", *IEEE Power and Energy Magazine*, vol. 13, no. 5, pp. 41–48, Sept. 2015.
[18] http://www.terna.it/en-gb/sistemaelettrico/dispacciamento/centronazionaledi controllo.aspx
[19] J. Pierre, D.J. Trudnowski, and M.K. Donnelly, "Initial results in electromechanical mode identification from ambient data," *IEEE Transactions on Power Systems*, vol. 12, no. 3, pp. 1245–1251, Aug. 1997.
[20] J.F. Hauer, C. J. Demeure, and L.L. Scharf, "Initial results in prony analysis of power system response signals," *IEEE Transactions on Power Systems*, vol. 5, no. 1, pp. 80–89, Feb. 1990.
[21] N. Zhou, J. Pierre, and J. Hauer, "Initial results in power system identification from injected probing signals using a subspace method," *IEEE Transaction on Power Systems*, vol. 21, no. 3, pp. 1296–1302, Aug. 2006.
[22] L. Vanfretti, S. Bengtsoon, and J.O Gjerde, "Preprocessing synchronized phasor measurement data for spectral analysis of electromechanical oscillations in the Nordic Grid," *International Transactions on Electrical Energy Systems*, early view at http://onlinelibrary.wiley.com/doi/10.1002/etep.1847/abstract.
[23] D. Trudnowski, J. Pierre, N. Zhou, *et al.*, "Performance of three mode-meter block-processing algorithms for automated dynamic stability assessment," *IEEE Transactions on Power Systems*, vol. 23, no. 2, pp. 680–690, May 2008.
[24] L. Ljung, *System Identification Theory for the User*. Upper Saddle River, New Jersey: Prentice Hall; 1999.

[25] N. Zhou, J.W. Pierre, D. Trudnowski, and R. Guttromson, "Robust RLS methods for on-line estimation of power system electromechanical modes," *IEEE Transactions on Power Systems*; vol. 22, no. 3, pp. 1240–1249, May 2007.

[26] P. Van Overschee and B. De Moor, *Subspace Identification for Linear Systems: Theory-Implementation-Applications*. London: Kluwer Academic Publishers; 1996.

[27] B. Kovacevic, M. Milosavljevic, and M. Veinovic, "Robust recursive AR speech analysis," *Signal Processing*, vol. 44, no. 2, pp. 125–138, June 1995.

[28] L. Vanfretti, S. Bengtsoon, and J.O Gjerde, "Preprocessing synchronized phasor measurement data for spectral analysis of electromechanical oscillations in the Nordic Grid," *International Transactions on Electrical Energy Systems*, vol. 25, no. 2, pp. 348–358, Feb. 2015.

Chapter 2

Reliability-based substation monitoring systems placement

Oscar Gomez[1] *and George J. Anders*[2]

Since the invention of phasor measurement unit (PMU), there has been growing interest in developing methodologies for finding the minimum number of units for complete system observability. The methods for the PMU placement must consider the fact that the network topology may change when the power system is affected by a contingency event (transmission lines and transformers). In order to design a robust wide area measurement (monitoring) systems (WAMS) which can ensure complete system observability under the failure of any transmission line or even a PMU, some works have considered power system contingencies and loss of measurements (failure of a PMU or its communication links).

For instance, in Reference 1, the authors presented a method for optimal placement of PMUs that ensures system observability under a prespecified number of critical contingencies, which are identified by performing beforehand a voltage stability analysis. Although these contingencies are critical for the system stability, they could have small probability of occurrence; therefore, contingencies with higher probability of occurrence and highly negative effect on the system observability could be omitted. In Reference 2, a method for the optimal placement of PMUs that considers two types of contingencies (single measurement loss and single-branch outage) was presented. The methodology uses a sequential addition approach to search for necessary candidates for single measurement loss and single-branch outage conditions, which are optimized by binary integer programming and a heuristic technique. In Reference 3, the integer quadratic programming approach was used to minimize the total number of PMUs under an outage of a single transmission line or one PMU; however, a list of branch outages to be considered is prepared beforehand. This model, which was based on numerical observability analysis, is computationally expensive. In Reference 4, an optimal set of PMUs, which maximize the measurement redundancy, was found using a nondominated sorting genetic algorithm and topological observability. The algorithm starts with a set of PMUs that ensures complete observability of the system,

[1]School of Electrical Technology, Universidad Tecnológica de Pereira, Pereira, Colombia
[2]Lodz University of Technology, Poland

and the additional PMUs are added in an iterative way until a predefined measurement redundancy has been achieved. In Reference 5, integer linear programming was proposed for solving the optimal placement of PMUs anticipating the loss of a PMU or a line outage. The effect of a single line outage is added directly to the model by using auxiliary variables. A technique for placing the PMUs in multiple stages over a given time period that ensures complete power system observability even under a branch outage or a PMU failure was presented in Reference 6. Reference 7 presents a multiobjective mathematical model for the PMU placement that minimizes the number of PMUs and maximizes the measurement redundancy.

The methodologies proposed in References 1–7 do not consider the stochastic nature of both WAMS and power system components, like transmission lines, which are essential for the proper operation of the monitoring system.

Although the monitoring system may be robust enough to maintain the system observability anticipating all possible contingencies, the number of PMUs could be very high, and the implementation of the system monitoring would be expensive. On the other hand, the random nature of contingencies causes that some transmission lines have higher probability of failure than others. Therefore, it is advisable to design a methodology that considers the random nature of the transmission line outages and WAMS component failures.

PMU placement considering random operating scenarios and random topologies was initially proposed in Reference 8. Authors proposed a methodology to find the optimal location of PMUs for wide-area monitoring and control of large disturbances; the methodology places a minimum number of PMUs that maximizes the useful information to monitor the system dynamic performance. In Reference 9, authors find the optimal number of PMUs to enhance the system observability by considering random component outages and PMU reliability. Through an iterative process, authors find the probability of observability associated with all buses, which are averaged to get a system index. This index is posteriorly used to select the best solution from all the possible ones. Although authors consider random outages of the WAMS components, and analytical reliability evaluation methods to calculate the probability of observability, the algorithm requires finding all the optimal solutions of the PMU placement problem, which might be very large for relatively large-scale systems with thousands of buses. The approach proposed in this chapter avoids finding all optimal solutions, it defines the WAMS reliability as the probability of observing all the buses under single component contingencies, and it finds the optimal solution without an exhaustive search of the possible PMU placements. The same authors, in Reference 10, developed a methodology for the staged PMU placement in a multiyear planning horizon where the average probability of observability is maximized at the intermediate planning stages. Initially, they calculate the minimum number of PMUs for the last stage; then, the multistage scheduling is determined in the context of the final solution by solving a subsidiary optimization model for each stage. The minimum number of PMUs for the final stage must satisfy the criterion that the probability of observability associated with all buses would be equal to or greater than a set of values that must be estimated or calculated a priori. The model for the probability of observability is a nonlinear function which must be linearized. The presented proposal in this chapter is based on the system state space, so all relations are already linear, and an integer linear programming model can be applied. Finally, authors in Reference 11 proposed a method for the optimal placement of PMUs and phasor data concentrators (PDCs) in local networks of a WAMS; the method minimizes the probability of failures in data transmission from PMUs to PDCs.

The methodologies proposed in References 8–11 considered the PMU reliability, but they did not consider the reliability of the components that form the monitoring system at the substation, that is, they assigned to each substation the same value of PMU reliability without considering the topology of the substation, the size, or the type of PMU. Because the same reliability value for all the buses was used, results were dominated by the objective of making the WAMS robust under transmission lines contingencies. Therefore, this chapter introduces the term substation monitoring system (SMS), which includes four basic components: PMUs, PDCs, communication network at the substation, and PMU application software. This term will be exposed in Section 1.1, and our objective will be to select the system buses where the SMSs must be installed to ensure full observability of the system.

In this chapter, we present the results of References 12 and 13, which introduced a new reliability-based model for the contingency-constrained SMSs placement. Initially, the probability of failure of the transmission lines is considered for placing the SMSs at the most reliable buses. Reliability of a bus for the monitoring system is defined as the availability of the basic components for the operation of the SMS and the availability of the adjacent lines to a bus, in such a way that the monitoring system can correctly observe its neighborhood buses. Therefore, the transmission lines probability of failure is considered during the selection of the minimal number of monitoring systems, and their location to monitor the system under normal operation and the most likely contingencies.

A probabilistic index will be presented to select the level of reliability that the system operator wants its WAMS to meet. The problem is formulated and solved using a binary integer linear programming model. Thus, the model simultaneously considers the reliability of the buses, lines and WAMS, as well as allowing for only some important contingencies to be considered by an introduction of the desired system reliability index. The zero-injection effect will not be considered because our objective is to improve the WAMS reliability, which is degraded when many components have to be available to make one zero-injection effect applicable. Because the effect of zero-injection will be excluded in the monitoring system placement process, each substation is made observable either by its own monitoring system or by monitoring systems at the adjacent buses.

The first step in evaluating the reliability of WAMS is to assess the SMS reliability based on the availability of its basic components. However, to the best of our knowledge, the reliability analysis has been based just on the PMU. Thus, authors in Reference 14 presented a hierarchical Markov modeling technique for reliability evaluation of phasor measurement unit. The PMU was divided into seven modules for reliability modeling in terms of its internal components. Reference 15 proposed a reliability model of the PMU that considers the PMU hardware; then, the Markov process was employed to analyze the proposed model and to obtain an equivalent two-state model of the PMUs. Reference 16 is an extension of Reference 14 where authors addressed the uncertainties of reliability data in PMU. Reference 17 proposed a reliability evaluation method based on Monte Carlo dynamic fault tree analysis (FTA) to conduct the reliability evaluation of a PMU. The PMU reliability model was constructed using dynamic fault tree modeling and analyzed using Monte Carlo simulation to evaluate the reliability indexes.

The approaches exposed in References 14–17 focused on the reliability evaluation of a PMU based on its basic components. The PMU was divided into several modules, and the reliability analysis for each subcomponent was developed individually. Then, the models for subcomponents were combined to evaluate the reliability of the entire unit.

Others works have developed methodologies to evaluate the WAMS reliability as a whole [18] and [19], or just a section of the WAMS like its communication system [20] and [21].

On the other hand, previous research papers have assumed PMUs with multiple channels. These PMUs could be placed at a bus and would provide bus voltage as well as current phasors along all branches incident to the bus. However, there is another class of PMUs known as branch PMUs. These PMUs are designed to monitor a single branch by measuring the voltage and current phasors at one end of the monitored branch, and its use will increase the reliability of the monitoring system at the substation.

This chapter also proposes an application of the FTA for the quantitative reliability evaluation of SMSs based on branch PMUs. The reliability evaluation is performed considering the PMU and all the additional equipment that are part of the SMS. Results obtained from the application of FTA are used in the model that will find the most reliable substations where branch PMUs must be placed. Finally, four indexes will be computed to evaluate the robustness of any monitoring system placement solution.

2.1 Substation monitoring system

2.1.1 Monitoring system components

An SMS includes four basic components: PMUs, PDCs, communication network, and PMU application software [22]. Figure 2.1 illustrates the layer representation with the basic components of an SMS.

The first layer represents the interface between the power system and the monitoring system, and it includes current and potential transformer (CT and PT), and the wiring to connect the instrument transformers to the PMUs associated with the substation bus bars or power lines. The second layer contains the substation PDC which collects measurements from PMUs to aggregate and sort them based on the time tag. The third layer refers to the algorithms and methodologies developed

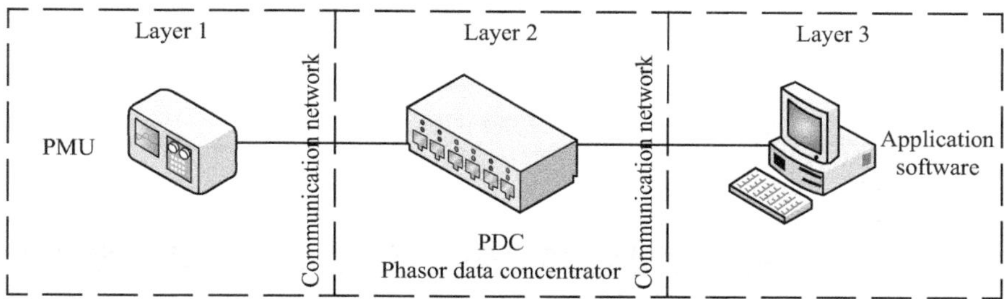

Figure 2.1 Components of SMS

for taking decisions. Finally, the information transaction between these three layers is realized by the communication network.

2.1.2 Branch PMU

Usually, it is assumed that PMUs have infinite number of channels to monitor phasor currents of all branches that are incident to the substation where the unit is installed; however, manufacturers produce PMUs with a limited number of channels to measure currents and voltages, so assuming unlimited number of channels is unrealistic.

On the other hand, there is another class of devices known as branch PMUs. These PMUs are designed to monitor a single branch by measuring the voltage and current phasors at one end of the monitored branch. Branch PMUs offer benefits like uniform distribution in the network, adaptability to deployment in multiple stages, higher reliability, and increasing the reliability of the monitoring system at the substation [23].

The reliability analysis conducted in this chapter considers that the SMS is based on branch PMUs. Figure 2.2 shows the location of the branch PMUs in a single busbar substation with three transmission lines.

2.1.3 Substation configuration

Most substations at the transmission system are organized similarly to one of four standard configurations, which are referred to as busbar configurations. Although there are some variations, the basic four busbar configurations are breaker-and-a-half, double-bus-double-breaker, double-bus-single-breaker, and ring bus.

Regardless of the substation configuration, the SMS must have the three layers shown in Figure 2.1. As we are considering branch PMUs, each outgoing line must have a set of CTs and PTs to connect the device. Figure 2.3 shows the components of layer one for two substations with different configurations (breaker-and-a-half and double-bus-double-breaker). We can observe that the number of components in layer 1 of the monitoring system depends on the number of transmission lines; it does not depend on the substation configuration. Therefore, the reliability analysis at each substation will be performed taking into account the number of transmission lines outgoing from the substation.

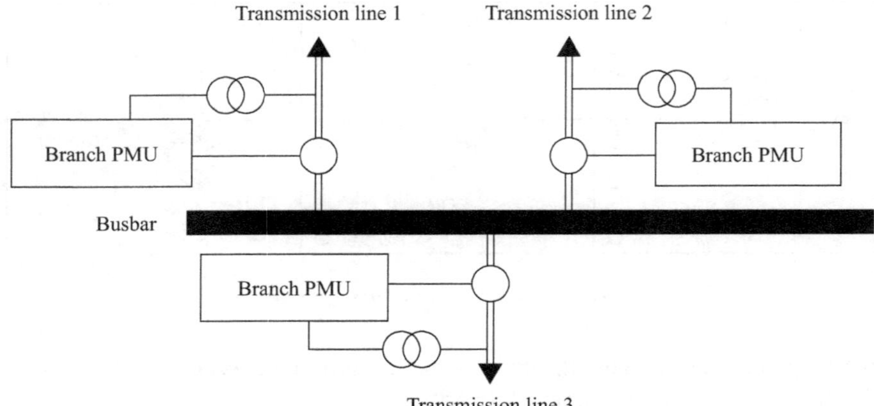

Figure 2.2 Branch PMU location at a single busbar substation

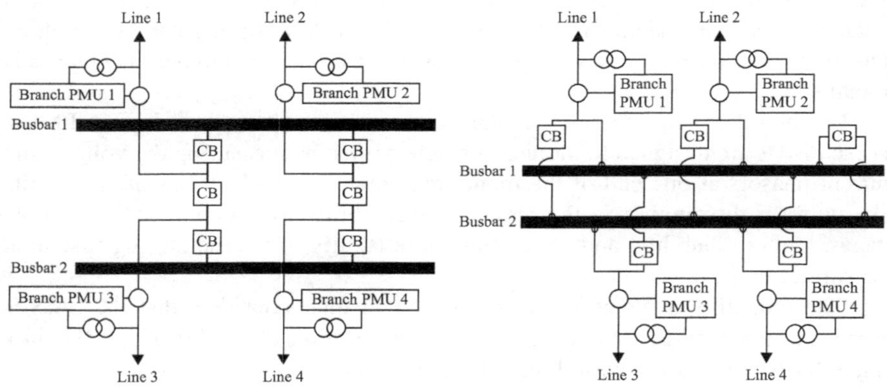

Figure 2.3 Layer 1 for two substations with different configurations (breaker-and-a-half and double-bus-double-breaker)

2.1.4 Substation communication system

The communication system can be divided into local and interstation communication systems. The local communication consists of links between the instrument transformers, PMUs, and PDC at the substation. The local communication is responsible for collecting and carrying the information to the PDCs. This information will be sent posteriorly to regional or national control centers through the wide area communication system.

In our analysis, only the local communication system of the substation will be considered. It will be modeled as a block in the FTA, and its probability of failure can be computed according to methodologies discussed in References 20 and 21.

The external communication system is usually designed to be highly reliable, so it will not be considered in our analysis.

2.2 Substation monitoring system reliability

The first step in designing a reliable WAMS is to assess the monitoring system reliability at each substation; it will be done by using FTA, which is a deductive failure analysis where an undesired state of the system is analyzed using Boolean logic to combine a group of lower-level events.

The objective of the SMS is to collect the electrical measurements and sending them to the regional or national control center; therefore, the undesired effect (the top event) is defined as losing observability of the substation. The situations that could cause losing observability of the substation are impossibility of getting voltage or current measurements, failure of the PDC, or failure of the local communication system. Figure 2.4 shows the FTA model of the described events.

The probability of the top event is calculated as:

$$Q_{\text{sub}_i} = 1 - ((1 - Q_{\text{mea}}) \cdot (1 - Q_{\text{pdc}}) \cdot (1 - Q_{\text{com}})) \tag{2.1}$$

where Q_{sub_i} is the failure probability of the monitoring system at substation *i*, Q_{mea} is the failure probability of the measurements, Q_{pdc} is the failure probability of the PDC, and Q_{com} is the failure probability of the internal communication system.

There is a measurement failure when it is not possible to measure the voltage or the current in the substation. It is defined by:

$$Q_{\text{mea}} = 1 - ((1 - Q_{\text{vm}}) \cdot (1 - Q_{\text{cm}})) \tag{2.2}$$

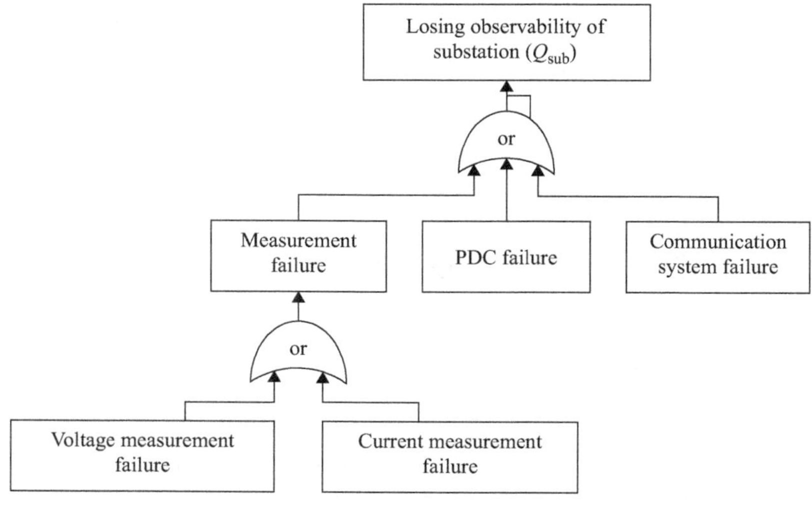

Figure 2.4 FTA model of an SMS

where Q_{vm} is the failure probability of the voltage measurement, and Q_{cm} is the failure probability of the current measurement.

When PMUs with several channels are used and a fault occurs in the PMU, this fault will affect the entire SMS; however, in case of branch PMU, the failure of one branch PMU will not affect the measurement of the voltage or current. Therefore, the failure probability of voltage and current measurement can be calculated as described below.

2.2.1 Failure probability of voltage measurement

Branch PMUs are designed to measure the voltage and current phasors at one end of the monitored branch; therefore, each branch PMU installed at the substation can measure the voltage phasor. Figure 2.5 shows the FTA model of the voltage measurement reliability assessment.

Because each branch PMU can measure the voltage at the substation, this measurement fails when all the measurements at the line bays fail. Therefore, the failure probability of the voltage measurement can be calculated by:

$$Q_{\mathrm{vm}} = 1 - \prod_{j=1}^{N_{\mathrm{lb}}} \left(1 - Q_{v_{\mathrm{bay}_j}}\right) \tag{2.3}$$

where Q_{vm} is the failure probability of the voltage measurement, $Q_{v_{\mathrm{bay}_j}}$ is the failure probability of the voltage measurement at the line bay j, and N_{lb} is the number of outgoing line bays in the substation.

$Q_{v_{\mathrm{bay}_j}}$ can be calculated as:

$$Q_{v_{\mathrm{bay}_j}} = 1 - \left(\left(1 - Q_{\mathrm{pt}_j}\right) \cdot \left(1 - Q_{\mathrm{bpmu}_j}\right) \cdot \left(1 - Q_{\mathrm{plink}_j}\right)\right) \tag{2.4}$$

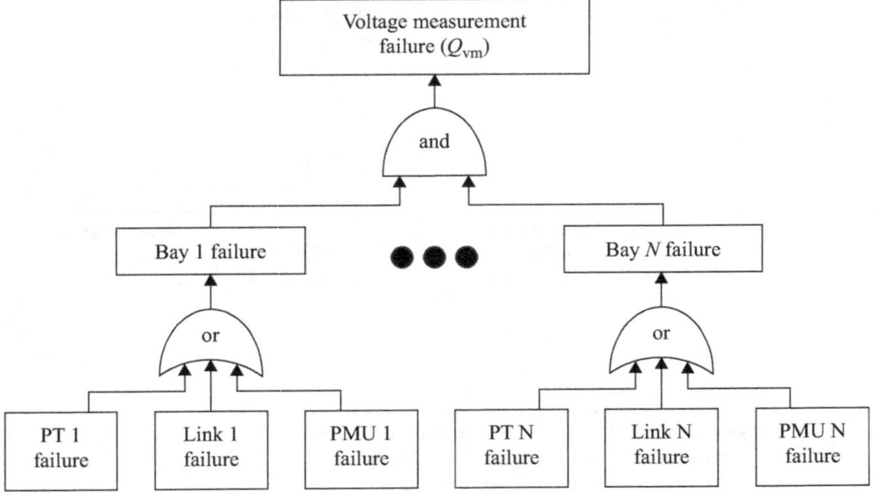

Figure 2.5 FTA model of voltage measurement

where Q_{pt_j} is the failure probability of PT at line bay j, Q_{bpmu_j} is the failure probability of the branch PMU at line bay j, and Q_{plink_j} is the failure probability of link between the instrument transformers and the branch PMU.

2.2.2 Failure probability of current measurement

The current measurement cannot be obtained when two or more measurements at the line bays fail. If one branch PMU fails, the current in the transmission line, where the failed branch PMU is located, can be estimated in the PDC using the other current phasors and the Kirchhoff laws. Thus, the failure probability of the current measurement can be determined by calculating the probability that two or more of N_{lb} current measurements fail. The current measurement at a line bay could fail if the CT, the branch PMU, or the link between the instrument transformers and the branch PMU fail (2.5).

$$Q_{\mathrm{cbay}_j} = 1 - \left(\left(1 - Q_{\mathrm{ct}_j}\right) \cdot \left(1 - Q_{\mathrm{bpmu}_j}\right) \cdot \left(1 - Q_{\mathrm{clink}_j}\right)\right) \tag{2.5}$$

where Q_{cbay_j} is the failure probability of the current measurement at the line bay j, Q_{ct_j} is the failure probability of CT at line bay j, Q_{bpmu_j} is the failure probability of the branch PMU at line bay j, and Q_{clink_j} is the failure probability of link between the instrument transformers and the branch PMU.

Figure 2.6 presents the FTA model of the current measurement. It is important to point out that the branch PMU would require three phase values for calculating the positive sequence, so the outage of one PT or CT in one of three phases would render the PMU unable to measure the voltage phasor.

Therefore, when we state that branch PMU, CT, PT or link between instrument transformers and the PMU fails, we refer to the set of elements required to do the measurements in the three phases.

On the other hand, the PDC is a fundamental component required to send the information to the regional or national control center; in consequence, the failure of this component is modeled as a block connected in series to the measurement failure block and the local communication system block (Figure 2.4), which is also

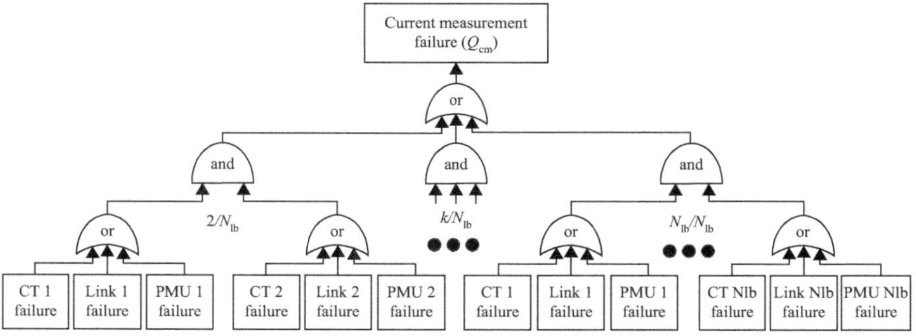

Figure 2.6 FTA model of current measurement

an essential element that allows transmission of the data between the monitoring system layers.

2.3 SMS placement based on bus reliability

The first step in evaluating the reliability of WAMS is to assess the SMS reliability based on the availability of its basic components. However, from the standpoint of the location in the power system, not only the availability of the SMS but also the availabilities of the transmission lines are necessary for the proper operation of the WAMS. In fact, transmission lines are indispensable for observing the neighboring buses of a substation with an installed monitoring system. Therefore, in order to select the most reliable substations, transmission line failures must be considered as a crucial element for the mission of the monitoring system.

Transmission line failures will be modeled as an independent single outage, that is, only the first-order contingencies will be considered, which is not related in terms of its cause to any other failures that may occur at the same time. According to the number of circuits between buses, transmission line failures can be modeled as a single component which can have either in service (up) or out of service (down) (Figure 2.7), or as a two (or more) repairable independent components with independent and common mode outages (Figure 2.8).

λ and μ denote the failure and repair rates, respectively, and the probability that the transmission line is in service (probability that the line is operating), is given by:

$$R = \frac{\mu}{\lambda + \mu} \tag{2.6}$$

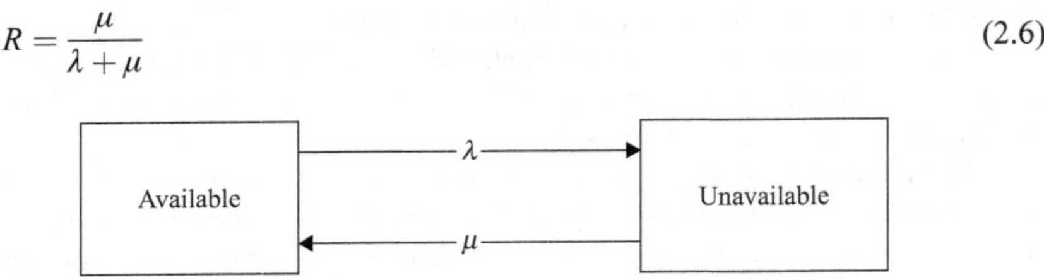

Figure 2.7 Model of a transmission line as a single component

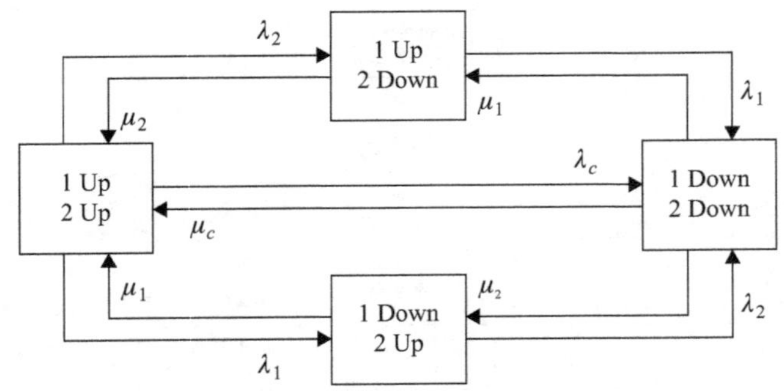

Figure 2.8 Model of a transmission line as a two component

Multiple circuit lines involved in a common-mode outage can still fail independently without influencing the failure of other components. When a common mode outage is concerned, the involved components are considered as a single entity, that is, multiple circuit lines connecting the same two buses are treated as a single connection with the equivalent reliability parameters λ_c and μ_c.

We will start by introducing an index representing the probability that the SMS at bus j and all of its adjacent lines are available:

$$P_{\text{in}_j} = R_{\text{sub}_j} \cdot \prod_{k=1}^{\Omega j} R_k \tag{2.7}$$

R_{sub_j} is the SMS reliability, R_k is the reliability or availability of the kth transmission line, and Ω_j is the set of adjacent transmission lines at bus j. The WAMS reliability for accomplishing its mission can be increased if the number of transmission lines that the SMS requires for observability is reduced, and the SMS is located at a bus where R_{sub} is the highest. That is, selecting the buses where SMS have lower unavailability ($Q_{\text{sub}} = 1 - R_{\text{sub}}$), and considering the most probable contingencies, it is possible to improve the observability of the system.

2.4 Reliability-based substation monitoring system placement considering transmission line outages

The SMS placement must be robust enough to maintain system observability anticipating possible transmission line outages; therefore, a reliability-based linear optimization model for the SMS placement should be developed in order to identify the most credible contingencies to be taken into account for the SMS location, that is, contingencies that have relatively high probability.

The proposed model strategically places an optimal number of SMSs to monitor the power system under normal operation and the single component contingencies that have high probability. In order to identify the most probable contingencies, state space enumeration technique is used.

This method involves defining all possible single component contingencies states of the system based on the states of the transmission lines. A state is defined by listing the successful and failed transmission lines in the system. The states that result in successful WAMS operation are identified, and the probability of occurrence of each successful state is calculated. The reliability of the WAMS can be calculated as the sum of all the successful state probabilities.

The probability associated with the normal state of the power system (system without contingencies) is calculated by:

$$P_{\text{ns}} = \prod_{k=1}^{N_{\text{L}}} R_k \tag{2.8}$$

where P_{ns} is the probability of the system normal state, N_{L} is the number of transmission lines, and R_k is the reliability (availability) of the transmission line k. If the system is observable under normal state, the WAMS reliability is P_{ns}.

On the other hand, the probability associated with any single component contingency state k can be calculated as:

$$P_{cs_k} = Q_k \cdot \prod_{i=1, i \neq k}^{N_L} R_i \tag{2.9}$$

where P_{cs_k} is the probability of the kth contingency state, and Q_k is the unreliability (unavailability) of the transmission line k.

If the system is observable under normal state, and it is still observable under the kth contingency state, then the WAMS is robust for contingency k^{th}, and the WAMS reliability based on this contingency can be computed as $P_{ns} + P_{cs_k}$.

With a view to enhancing the WAMS reliability, an optimization model can be proposed to include the maximization of the reliability criterion. First, the following objective function is proposed:

$$\min \left[\sum_{i=1}^{N} (1 + Q_{sub_i}) \cdot x_i + \sum_{k=1}^{N_L} -P_{cs_k} \cdot y_k \right] \tag{2.10}$$

where x_i is a binary decision variable that represents the placement of an SMS at bus i, y_k is a binary decision variable that represents observability of the system under the kth contingency state (single line outage k). In other words, $y_k = 1$ means that contingency of line k does not impact the system observability; by contrast $y_k = 0$ means that complete system observability is not reached under this contingency.

The main objective of the SMS placement problem is to reduce the number of SMSs, so it is a minimization problem; therefore, P_{cs_k} is introduced with a negative sign. The objective function will be minimized if more y_k are equal to one; as a result, the solution will include the single component contingency states with the highest values for P_{cs_k}.

Additionally, the following two constraints are considered for each line k, one associated with the sending and one for the receiving bus of this line.

$$\begin{aligned} f_{k_s}(x, y) &= \sum_{j \in \Omega_{sb}} x_j - y_k \geq 0 \\ f_{k_r}(x, y) &= \sum_{j \in \Omega_{rb}} x_j - y_k \geq 0 \end{aligned} \tag{2.11}$$

where Ω_{sb} is the set of buses composed of the sending bus and the buses connected to the sending bus of the transmission line k, and Ω_{rb} is the set of buses composed of the receiving bus and the buses connected to the receiving node of the transmission line k.

If the system is fully observable under the contingency y_k, the constraints (2.11) ensure that an SMS must be placed either at the sending bus of line k or at any adjacent bus; also, an SMS must be placed at the receiving bus of line k or at any adjacent bus.

Finally, the model for the reliability-based SMS placement considering transmission line outages is:

$$
\begin{aligned}
&\min\left[\sum_{i=1}^{N}(1+Q_{\text{sub}_i})\cdot x_i+\sum_{k=1}^{N_L}-P_{\text{cs}_k}\cdot y_k\right]\\
&f_j(x)=\sum_{i\in\Omega j}x_i\geq 1\\
&f_{k_s}(x,y)=\sum_{j\in\Omega_{\text{sb}}}x_j-y_k\geq 0\\
&f_{k_r}(x,y)=\sum_{j\in\Omega_{\text{rb}}}x_j-y_k\geq 0
\end{aligned}
\tag{2.12}
$$

The objective function minimizes the number of SMSs, and it maximizes the number of contingencies covered by the WAMS. Because P_{cs_k} is very small ($P_{\text{cs}_k} << 1$), the minimization of the number of SMSs has a high priority, and the maximization of the number of covered contingencies does not increase the number of required SMSs. Additionally, contingencies that may have lower probability of occurrence but high consequences for the WAMS could also be incorporated in the model multiplying $P_{\text{cs}_k}$by C_k, which is a number greater than 1 and represents the importance of contingency *k*.

In order to achieve a desired level of reliability, (R_{wd}) of the WAMS, a new constraint (2.13) is introduced as:

$$
\sum_{k=1}^{N_L}P_{\text{cs}_k}\cdot y_k+P_{\text{ns}}\geq R_{\text{wd}} \tag{2.13}
$$

Here, the improvement in the WAMS reliability is defined as the reduction of the probability of losing observability due to contingencies in the power system, increasing the observed states. Hence, a high R_{wd} implies that system will be observable under several single component contingencies (high-probability contingencies). As R_{wd} decreases, the system will not be observable under contingencies with low probability. Therefore, the R_{wd} is the summation of the probabilities of the observable states by the WAMS.

2.5 Evaluation of reliability-based SMS placement

Four indexes will be computed to evaluate the robustness of any monitoring system placement solution. The first two indexes are based on the reliability of the SMSs and the single component contingencies considered for observability by the WAMS; the third index is denominated average probability of observability (APO) proposed in Reference 9, and the fourth is the failure probability of WAMS caused by a failure in a substation or a single line contingency not considered for observability.

A contingency considered for observability means that the power system is still observable under this contingency; on the contrary, a contingency not considered for observability refers to the case where a transmission line or branch contingency causes loss of observability.

The failure probability of the WAMS based on the unreliability of the substations ($Q_{\text{wams}_{\text{sub}}}$) is evaluated by using:

$$Q_{\text{wams}_{\text{sub}}} = 1 - \prod_{i=1}^{N_{\text{ms}}} \left(1 - Q_{\text{sub}_i}\right) \tag{2.14}$$

where N_{ms} is the number of substations where the monitoring system will be installed.

The WAMS reliability based on single component contingencies ($R_{\text{wams}_{\text{line}}}$) will be calculated as:

$$R_{\text{wams}_{\text{line}}} = P_{\text{ns}} + \sum_{k=1}^{N_{\text{c}}} P_{\text{cs}_k} \tag{2.15}$$

where N_{c} is the number of contingencies considered for observability.

The best solution must minimize the failure probability of the WAMS based on the reliability of the substations while maximizing the reliability based on single component contingencies.

The APO [9] is defined as:

$$\text{APO} = \frac{1}{N} \cdot \sum_{i=1}^{N} \text{PO}_i \tag{2.16}$$

where N is the number of buses (substations), and PO_i is the probability of observability at bus i. PO is computed at each bus of the system; therefore, the buses are separated into buses observable through its SMS, and buses observable through monitoring systems located at incident substations.

For the case of buses observable through monitoring systems at incident buses, the probability of observability is:

$$\text{PO}_i = 1 - \prod_{i \in \Omega_{\text{b}}} \left(1 - A_{ij}\right) \tag{2.17}$$

where b is the set of buses with monitoring systems incident to the bus i, and A_{ij} is defined as:

$$A_{ij} = \left(1 - Q_{\text{sub}_j}\right) \cdot R_{ij} \tag{2.18}$$

In (2.18), Q_{sub_j} is the SMS unavailability at the incident bus j, and R_{ij} is the availability of transmission line ij.

For the case of buses observable through its monitoring system, the probability of observability is:

$$\text{PO}_i = 1 - Q_{\text{sub}_j} \tag{2.19}$$

According to Reference 9, the PMU placement solution with the largest APO should be the best solution. Although the PO considers the failure probability in the transmission lines and the SMSs, PO indexes are averaged to get the APO index, that is, PO of substations are combined linearly, which is not related with the probabilistic criteria.

Finally, the failure probability of WAMS caused by a failure of an SMS or a single component contingency not considered for observability is calculated by:

$$Q_{\text{wams}} = 1 - \left(\prod_{i=1}^{N_{\text{ms}}} (1 - Q_{\text{sub}_i}) \cdot \prod_{k=1}^{N_{\text{uc}}} R_k \right) \tag{2.20}$$

where Q_{wams} is the failure probability of the WAMS, and N_{uc} is the number of single component contingencies not considered.

2.6 Numerical studies

The Western System Coordinating Council (WSCC) 3-machine, 9-bus system and the IEEE 57-bus test systems are used to demonstrate the proposed methodology. The topologies are shown in Figures 2.9 and 2.10. Initially, the optimization model will be used in the traditional way, considering the same value of SMS reliability for all the buses of the system. It was defined as 0.995498, which is reported in References 2 and 6 as the reliability of one PMU. The availabilities of the transmission lines for the IEEE 9-bus test and the IEEE 57-bus are provided in References 9 and 10, respectively. Then, the SMS reliability at each bus of the system will be calculated to be used in the proposed optimization model.

The simulations were carried out on a Toshiba laptop Intel CORE i3 2.27-GHz, and the integer programming problem was solved using the TOMLAB Optimization Toolbox [24]. The required CPU times show that the proposed optimization model can be applied to real-world networks.

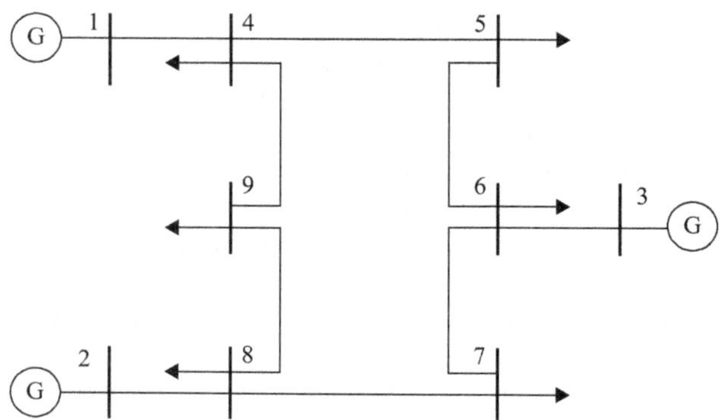

Figure 2.9 WSCC 3-machine, 9-bus system

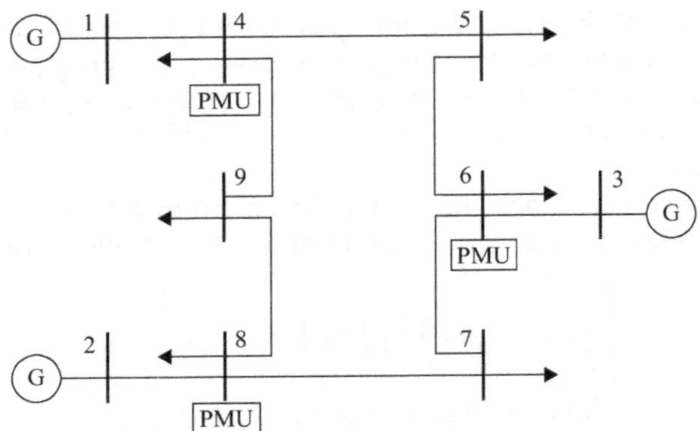

Figure 2.10 PMU location for the WSCC 3-machine, 9-bus system

Table 2.1 Probability of N – 1 contingency states for the WSCC system

From	To	P_{cs_k}	From	To	P_{cs_k}	From	To	P_{cs_k}
1	4	0.0089	4	5	0.0046	6	7	0.0023
2	8	0.0033	4	9	0.0054	7	8	0.0053
3	6	0.0055	5	6	0.0091	8	9	0.0044

2.6.1 Reliability-based PMU placement in the WSCC 3-machine, 9-bus system

The probability of the system normal state is calculated as 94.99%, and the probabilities associated with the single component contingency states are shown in Table 2.1.

The optimization model (2.12) was solved in order to find the optimal number of PMUs required for full observability of the system and maximum reliability of the WAMS. Figure 2.11 shows the obtained location for the PMUs where three PMUs are assigned to buses 4, 6, and 8. Buses 1, 2, 3, 5, 7, and 9 are observable using connecting transmission lines.

The system is observable under normal state, and it is still observable under contingencies of lines 4-9, 8-9, 7-8, 6-7, 5-6 and 4-5; so, the WAMS reliability is equal to 98.10%. The obtained solution is equal to the one found in Reference 6, where the probability of observability is maximized.

If the constraint representing a desired level of reliability is considered in the model, and a reliability of 99% is fixed, the optimal placement of the PMUs is at buses 1, 4, 6, and 8. As explained earlier, this restriction can be used to improve the reliability of the WAMS resulting in a more robust system under the most probable contingencies. In the previous solution, outages of lines 1-4, 2-8, and 3-6 were not covered for the WAMS system. From Table 2.1, we observe that the contingency state corresponding to the outage of line 1-4 has the largest probability; therefore, the solution is to place a new PMU at bus 1 which helps to cover this contingency

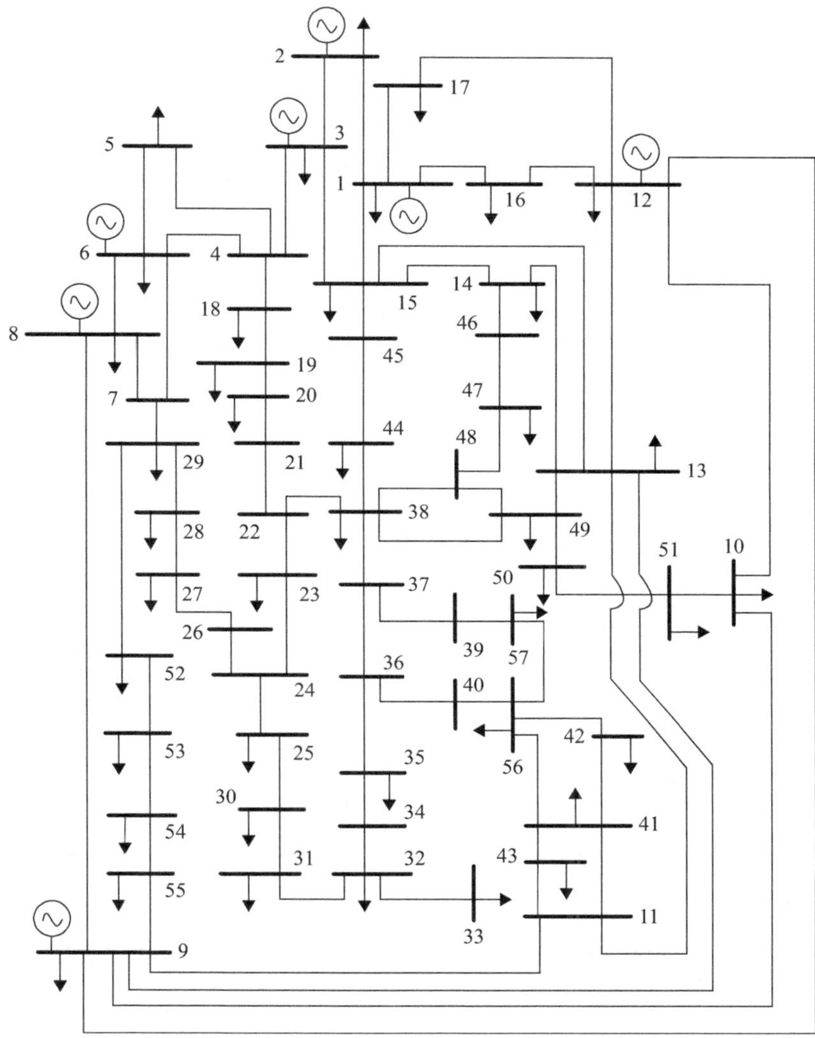

Figure 2.11 IEEE 57-bus test system

state. If an improvement in the reliability is required, more PMUs should be installed at buses that make the system observable under the most probable contingencies.

2.6.2 Reliability-based PMU placement in the IEEE 57-bus test system

We will illustrate the application of the model on the 57-bus network, which is fairly large but small enough to allow a visualization of the proposed solution.

Table 2.2 Probability of single component contingency states for the IEEE 57-bus test system

No.	From	To	P_{cs_k}	No.	From	To	P_{cs_k}	No.	From	To	P_{cs_k}
1	1	2	0.00274	28	14	15	0.00446	55	41	42	0.00219
2	2	3	0.00384	29	18	19	0.00556	57	41	43	0.00418
3	3	4	0.00164	30	19	20	0.00356	56	38	44	0.00501
4	4	5	0.00529	31	21	20	0.00260	58	15	45	0.00274
5	4	6	0.00515	32	21	22	0.00164	59	14	46	0.00473
6	6	7	0.00487	33	22	23	0.00391	60	46	47	0.00425
7	6	8	0.00232	34	23	24	0.00473	61	47	48	0.00116
8	8	9	0.00384	35	24	25	0.00308	62	48	49	0.00391
9	9	10	0.00123	36	24	25	0.00157	63	49	50	0.00287
10	9	11	0.00308	37	24	26	0.00232	64	50	51	0.00536
11	9	12	0.00260	38	26	27	0.00184	65	10	51	0.00116
12	9	13	0.00184	39	27	28	0.00239	66	13	49	0.00322
13	13	14	0.00198	40	28	29	0.00315	67	29	52	0.00308
14	13	15	0.00308	41	7	29	0.00356	68	52	53	0.00322
15	1	15	0.00157	42	25	30	0.00322	69	53	54	0.00467
16	1	16	0.00391	43	30	31	0.00515	70	54	55	0.00260
17	1	17	0.00329	44	31	32	0.00349	71	11	43	0.00473
18	3	15	0.00432	45	32	33	0.00549	72	44	45	0.00329
19	4	18	0.00432	46	34	32	0.00404	73	40	56	0.00130
20	4	18	0.00404	47	34	35	0.00556	74	56	41	0.00191
21	5	6	0.00130	48	35	36	0.00226	75	56	42	0.00446
22	7	8	0.00143	49	36	37	0.00425	76	39	57	0.00329
23	10	12	0.00260	50	37	38	0.00246	77	57	56	0.00184
24	11	13	0.00356	51	37	39	0.00418	78	38	49	0.00267
25	12	13	0.00411	52	36	40	0.00432	79	38	48	0.00391
26	12	16	0.00301	53	22	38	0.00143	80	9	55	0.00198
27	12	17	0.00487	54	11	41	0.00232				

Initially the probabilities associated with the single component contingency states (Table 2.2) are calculated.

Then, the optimization model (2.12) is solved to find the optimal number of PMUs required for full observability of the system, and to find the places that maximize the reliability of the WAMS. Seventeen PMUs are necessary to ensure complete observability of the system, and these are located as shown in the first row of Table 2.3.

The probability associated with the normal state (P_{ns}) in the test system is 68.14%, which was calculated by using (2.8). In the solution found by the model (2.12), which is shown in the first row of Table 2.3, the system is observable under fifty-one contingencies, so the WAMS reliability is 84.57%.

The maximum reliability that can be reached is 94.35% because it is the probability of the normal state and the single component contingency states. In order to get the higher level of reliability, N-2 and higher (N-k) contingencies should be modeled in the optimization problem; however, these are the

Table 2.3 Reliability-based PMU placement at the IEEE 57-bus test system

R_{wd} (%)	Number of PMUs	PMU location	Covered $N-1$ contingencies
84.57	17	1, 6, 9, 15, 18, 20, 24, 28, 30, 32, 36, 38, 41, 46, 50, 53, 57.	51
85	18	1, 4, 7, 9, 14, 19, 22, 24, 28, 30, 32, 36, 39, 41, 44, 47, 50, 53.	54
90	22	1, 4, 6, 12, 14, 19, 22, 24, 27, 29, 30, 32, 36, 39, 41, 44, 47, 49, 51, 53, 55, 56.	69
94	27	1, 3, 5, 8, 12, 14, 18, 20, 22, 24, 27, 29, 30, 32, 33, 34, 36, 39, 41, 43, 44, 47, 49, 51, 53, 55, 56.	79

contingencies that constitute 5.65% of all possible states (unlikely states), so it would be expensive to design a WAMS to account for all contingencies.

Table 2.3 shows the PMU location and the number of covered contingencies after solving the model (2.12) with the constraint (2.13) for several desired reliability values (R_{wd}).

As we can see from Table 2.3, when the desired level of reliability increases, the required number of PMUs also increases; therefore, the system is observable under a larger number of high probability contingencies.

2.6.3 Reliability analysis of the monitoring systems in the IEEE 9-bus test system

First, the failure probability of the SMS for each substation in the system will be evaluated. The reliabilities of PTs and CTs are 0.998542 and 0.999584, respectively, which are reported in Reference 25; the availability of the links between instrument transformers and the PMU is 0.999, and the branch PMU availability is 0.995497 according to Reference 15.

Table 2.4 shows the number of line bays and the SMS reliability for each system bus.

Four solutions were found by solving the classical integer linear programming model for the PMU placement. Each solution has three substations, where branch PMUs could be installed. Table 2.5 shows these solutions, and the values of $Q_{\mathrm{wams}_{\mathrm{sub}}}$, $R_{\mathrm{wams}_{\mathrm{line}}}$, APO, and Q_{wams} for each solution. In Table 2.5, node indicates the power system substation where the branch PMUs must be installed.

Solution 1 from Table 2.5 is Pareto-optimal because there is no other solution that is better for $R_{\mathrm{wams}_{\mathrm{line}}}$ and equal or superior with respect to $Q_{\mathrm{wams}_{\mathrm{sub}}}$; also, solution 1 has the lowest Q_{wams}. However, the APO index shows that solution 4 is the best as consequence of the linear relationship used to calculate the wide area index APO. Also, it is possible to observe that when the same value of reliability for all buses is used, results are dominated by the objective of making the WAMS robust under several contingencies. This can be seen in Section 6.1 where solution was to install SMSs at buses 4, 6, and 8.

Table 2.4 Number of line bays and substation monitoring system reliability

Bus	Line bays	Q_{sub}
1	1	0.016905
2	1	0.016905
3	1	0.016905
4	4	0.036809
5	3	0.030145
6	4	0.036809
7	3	0.030145
8	4	0.036809
9	3	0.030145

Table 2.5 Set of solutions of PMU placement problem for the 9-bus system

Sol.	Node	$Q_{wams_{sub}}$	$R_{wams_{line}}$	APO [%]	Q_{wams}
1	2, 4, 6	0.0879	0.9733	96.86	0.1121
2	3, 4, 8	0.0879	0.9712	96.83	0.1141
3	1, 6, 8	0.0879	0.9676	96.80	0.1174
4	4, 6, 8	0.1064	0.9811	97.28	0.1229

The solution found by model (2.12) was to install monitoring system at buses 2, 4, and 6, that is, it found the solution 1 in the set of solutions shown in Table 2.5. Therefore, the model (2.12) effectively finds the minimum number of substations with higher reliability while covering the most credible contingencies.

2.6.4 Reliability analysis of the monitoring systems in the IEEE 57-bus test system

The probability of the normal state of the system was calculated as 68.13%, and the probabilities associated with the single component contingency states were shown in Table 2.2. Table 2.6 shows the number of line bays, and the SMS reliability for each system bus.

Several optimal solutions were found by solving the classical integer linear programming model for the PMU placement. Seventeen is the optimal number of substation where the monitoring systems must be located. One thousand nine hundred seventeen optimal solutions were found, and they are shown in Figure 2.12.

Values of $Q_{wams_{sub}}$, $R_{wams_{line}}$, APO, and Q_{wams} for five representative solutions, which are part of the Pareto border, are presented in Table 2.7, where node indicates the power system substation where the branch PMUs must be installed.

Model (2.12) was solved for the 57-bus test system. The solution found by this model coincides with the solution number 3 from Table 2.7. If the monitoring

Table 2.6 Number of line bays and substation monitoring system reliability

Bus	Line bays	Q_{sub}	Bus	Line bays	Q_{sub}
1	5	0.04347	30	3	0.03014
2	3	0.03014	31	3	0.03014
3	4	0.03681	32	4	0.03681
4	5	0.04347	33	2	0.02343
5	3	0.03014	34	2	0.02343
6	5	0.04347	35	3	0.03014
7	3	0.03014	36	3	0.03014
8	4	0.03681	37	3	0.03014
9	7	0.05678	38	6	0.05013
10	4	0.03681	39	2	0.02343
11	4	0.03681	40	2	0.02343
12	6	0.05013	41	5	0.04347
13	7	0.05678	42	3	0.03014
14	4	0.03681	43	3	0.03014
15	6	0.05013	44	3	0.03014
16	3	0.03014	45	2	0.02343
17	3	0.03014	46	2	0.02343
18	4	0.03681	47	3	0.03014
19	3	0.03014	48	3	0.03014
20	3	0.03014	49	5	0.04347
21	2	0.02343	50	3	0.03014
22	3	0.03014	51	3	0.03014
23	3	0.03014	52	3	0.03014
24	4	0.03681	53	3	0.03014
25	4	0.03681	54	3	0.03014
26	2	0.02343	55	3	0.03014
27	3	0.03014	56	5	0.04347
28	3	0.03014	57	3	0.03014
29	4	0.03681			

systems are located at the 17 substations proposed by solution 3, the WAMS can operate under 46 single component contingencies.

Solution 3 has the highest average probability of observability (96.67%), and the lowest Q_{wams}; additionally, solution 3 is not dominated by any other solution with regard to $Q_{wams_{sub}}$ and $R_{wams_{line}}$. Therefore, the model (2.12), which converts the original problem with multiple objectives into a single-objective optimization problem, finds the minimum number of substations with highest reliability while covering the most credible contingencies.

2.7 Conclusions

This chapter has presented an application of the FTA to calculate the probability of failure of an SMS based on branch phasor measurement units.

SMSs based on branch PMU are more reliable than those based on PMUs with multiple channels. When a fault occurs in a PMU with larger number of channels, it

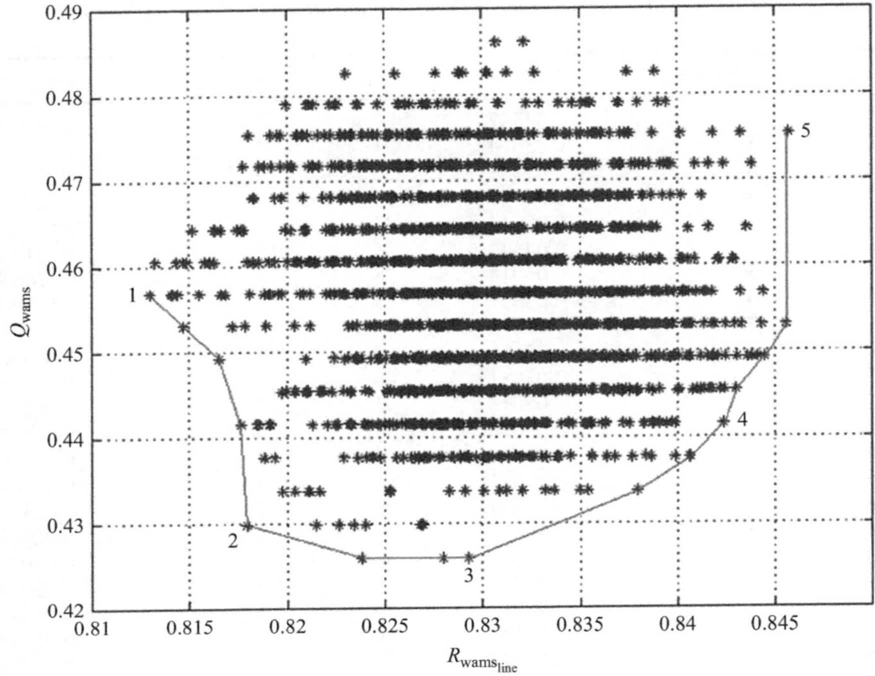

Figure 2.12 Multiple solutions of the PMU placement problem in the test system

Table 2.7 Set of solutions of PMU placement problem for the 57-bus system

Sol.	Node	$Q_{wams_{sub}}$	$R_{wams_{line}}$	APO[%]	Q_{wams}
1	1, 4, 9, 20, 24, 26, 29, 31, 32, 36, 38, 39, 41, 44, 46, 51, 54	0.45693	0.81296	96.56	0.55136
2	2, 6, 12, 19, 22, 25, 27, 32, 36, 39, 41, 45, 46, 47, 50, 52, 55	0.42989	0.81792	96.37	0.52561
3	2, 6, 12, 19, 22, 26, 29, 30, 32, 36, 39, 41, 45, 46, 48, 50, 54	0.42595	0.82928	96.67	0.51431
4	1, 4, 6, 10, 19, 22, 26, 29, 30, 32, 36, 39, 41, 45, 46, 49, 54	0.44157	0.84234	96.64	0.51842
5	1, 6, 9, 15, 18, 20, 24, 28, 30, 32, 36, 38, 41, 46, 50, 53, 57	0.47540	0.84571	96.45	0.54535

can affect the entire SMS; however, in case of branch PMU only the particular line bay would be affected.

The number of components of the SMS based on branch PMUs depends on the number of line bays or transmission lines; it does not depend on the substation configuration. Therefore, the reliability analysis at each substation can be performed according to the number of transmission lines outgoing from the substation. It allows the application of the methodology on any system with little information about the characteristics of each substation.

Failure probability of the current and voltage measurements can be separated and calculated independently. Branch PMU gives redundancy in the SMS because the failure of one device (TC, TP, link, or branch PMU) does not necessarily affect the measurements.

This chapter also has proposed an integer linear model for the optimal contingency-constrained PMU placement in electric power networks. The methodology considers the probability of failure of the power system components that could prevent normal operation of the SMS. The approach is based on selecting an appropriate number of SMSs to meet observability and reliability criteria considering single component contingencies.

The objective function of the classical optimization model was modified in order to find solutions that increase the availability of the measuring equipment. Therefore, the presented model locates the SMSs at buses that result in the best global reliability of the WAMS.

State space enumeration technique was used to evaluate the reliability of the wide area measurement system, which involves defining mutually exclusive states of a system based on the states of its components. A model for finding the optimal number of SMSs to monitor the system under normal operation, and the single component contingency states that have high probability of occurrence was developed. A linear constraint in the model, which considers the level of desired reliability of the WAMS (R_{wd}), was proposed. The model can be used in two ways, to enhance the reliability of an existing WAMS and to design a new WAMS with a prespecified value of reliability.

Results have shown that the proposed model effectively locates the minimum number of substations with higher reliability and includes the most credible contingencies. The model correctly converts the original problem with multiple objectives into a single-objective optimization problem, and it finds the non-dominated Pareto optimal solution.

References

[1] Sodhi R., Srivastava S., Singh S., "Optimal PMU placement to ensure system observability under contingencies". *Presented at IEEE Power and Energy Society General Meeting;* Calgary, Alberta, Canada, 2009.

[2] Rakpenthai C., Premrudeepreechacharn S., Uatrongjit S., Watson N., "An optimal PMU placement method against measurement loss and branch outage". *IEEE Transaction on Power Delivery*. 2007;22:101–107.

[3] Chakrabarti S., Kyriakides E., Eliades D., "Placement of synchronized measurements for power system observability". *IEEE Transaction on Power Delivery*. 2009;24:12–19.

[4] Milosevic B., Begovic M., "Non-dominated sorting genetic algorithm for optimal phasor measurement placement". *IEEE Transaction on Power Systems*. 2003;18:69–75.

[5] Aminifar F., Khodaei A., Fotuhi-Firuzabad M., Shahidehpour M., "Contingency-constrained PMU placement in power networks". *IEEE Transaction on Power Systems*. 2010;25:516–523.

[6] Sodhi R., Srivastava S., Singh S., "Multi-criteria decision-making approach for multistage optimal placement of phasor measurement units". *IET Generation, Transmission & Distribution*. 2011;5:181–190.

[7] Peng C., Sun H., Guo J., "Multi-objective optimal PMU placement using a non-dominated sorting differential evolution algorithm". *International Journal of Electrical Power & Energy Systems*. 2010;32:886–892.

[8] Kamwa I., Grondin R., "Pmu configuration for system dynamic performance measurement in large multiarea power systems". *IEEE Transaction on Power Systems*. 2002;17:385–394.

[9] Aminifar F., Fotuhi-Firuzabad M., Shahidehpour M., Khodaei A., "Observability enhancement by optimal PMU placement considering random power system outages". *Energy Systems*. 2011;2:45–65.

[10] Aminifar F., Fotuhi-Firuzabad M., Shahidehpour M., Khodaei A., "Probabilistic multi-stage PMU placement in electric power systems". *IEEE Transaction on Power Delivery*. 2011;26:841–849.

[11] Fesharaki F., Hooshmand R., Khodabakhshian A., "A new method for simultaneous optimal placement of PMUs and PDCs for maximizing data transmission reliability along with providing the power system observability". *Electric Power Systems Research*. 2013;100:43–54.

[12] Gomez O., Anders G., Rios M., "Reliability-based PMU placement in power systems considering transmission line outages and channel limits". *IET Generation, Transmission & Distribution*. 2014;8:121–130.

[13] Gomez O., Portilla C., Rios M., "Reliability analysis of substation monitoring systems based on branch PMUs". *IEEE Transactions on Power Systems*. 2015;30(2):962–969.

[14] Yang W., Li W., Jiping L., "Reliability analysis of phasor measurement unit using hierarchical markov modeling". *Electric Power Components and Systems*. 2009;37:517–532.

[15] Aminifar F., BagheriShouraki S., Fotuhi-Firuzabad M., Shahidehpour M., "Reliability modeling of PMUs using fuzzy sets". *IEEE Transaction on Power Delivery*. 2010;25:2384–2391.

[16] Wang Y., Li W., Zhang P., Wang P., Jiping L., "Reliability analysis of phasor measurement unit considering data uncertainty". *IEEE Transaction on Power Systems*. 2012;27:1503–1510.

[17] Peng Z., Ka W., "Reliability evaluation of phasor measurement unit using Monte Carlo dynamic fault tree method". *IEEE Transactions on Smart Grid*. 2012;3:1235–1243.

[18] Wen-xia L., Nian L., Yong-feng F., Li-xin Z., Xin Z., "Reliability analysis of wide area measurement system based on the centralized distributed

model". *Presented at Power Systems Conference and Exposition*; Seattle, WA, USA, 2009.

[19] Yang W., Li W., Jiping L., "Reliability analysis of wide-area measurement system". *IEEE Transaction on Power Delivery*. 2010;25:1483–1491.

[20] Dai Z., Wang Z., Jiao Y., "Reliability evaluation of the communication network in wide-area protection". *IEEE Transaction on Power Delivery*. 2011;26:2523–2530.

[21] Hai Y., Yue Y., Qigui Y., Haiqing Y., "Analysis on the reliability of wide area protection communication system". Proceedings of the IEEE 14th International Conference on Communication Technology; PyeongChang, Korea (South), Nov 2012. pp. 329–333.

[22] Zhu K., Chenine M., Konig J., Nordstrom L., "Data quality and reliability aspects of ict infrastructures for wide area monitoring and control systems". *Presented at International Conference on Critical Infrastructure*; Beijing, China, 2010.

[23] Emami R., Abur A., "Robust measurement design by placing synchronized phasor measurements on the network branches". *IEEE Transaction on Power Systems*. 2010;25:38–43.

[24] The TOMLAB optimization environment. Available from http://tomopt.com [Accessed 28 Oct 2014].

[25] Rice M., Heydt G., "The measurement outage table and state estimation". *IEEE Transaction on Power Systems*. 2010;23:353–360.

Chapter 3

System integrity protection scheme based on PMU technology

Srdjan Skok[1]

3.1 Introduction

The power transmission system (PTS) has to meet strict demands for greater reliability, with fewer investments made into the main infrastructure. By implementing new digital and adaptive solutions for protection and control of the entire PTS, a new generation of the transmission system will be created, which is widely known as intelligent power transmission system or smart transmission grid (STG).

Liberalization of electricity market, generation from distributed renewable energy sources (RES), and continuous growth in energy consumption set economic profit ahead of technological requirements, as a determining factor in strategic decisions of power system development. Consequences are reaching as far as fewer investments into construction of new facilities, especially of the transmission system, which again results in questionable protection and control, with lowered protection margins. This challenge requires new technological solutions that will be able to meet real-time monitoring, protection, and control demands.

The aim of many experts all over the world is research and development of an intelligent system for protection and control of the PTS, as well as its deployment in a section of the PTS. The intelligent system requires the creation of a system integrity protection scheme (SIPS) in case of failure. Also, adequate circuit architecture, necessary for the implementation of the developed system scheme, has to be created and deployed.

Wide area monitoring, protection and control (WAMPAC) is a concept that uses synchronized measurement technology (SMT) to counteract the propagation of large disturbances. Monitoring of wide area (WAMS) is the main function currently implemented in power systems, but with the development of protection and control schemes, WAMPAC is expected to improve security and reliability of power system operation, as well as disturbances mitigation and blackout prevention.

[1]Faculty of Engineering, University of Rijeka, Croatia

Although the idea of synchronized measurement is not new, its implementation is enabled with the use of time stamp from global positioning system (GPS). Also, implementation of the wide area schemes becomes possible due to the lower costs of technologies such as phasor measurement units (PMUs), phasor data concentrators (PDCs), and modern communication infrastructure.

Energy consumption increase and the modern trends in the electric power system (EPS) operation force the grid to operate closely to its stability limits with small power margins. Rapid development of electric power market, where often technical aspect is not emphasizes enough, but only economical one, treating electric power as any other merchandise, puts additional effort on the infrastructure which is not always upgraded or maintained as it should be. As a result we can witness huge electric power blackouts, some of them in the last decade. Power blackouts are low probability but extremely costly conditions, which frequency is increasing over years, due to the grid complexity and trends which comprise distributed renewable energy, two-way power flow, deregulation, liberalization, etc. Usually the triggering event is a contingency or a combination of few events. Such wide area disturbances cannot be treated or even "seen" by using local measurement and systems. For wide area monitoring, protection and control, and in the end maintaining power system integrity, information about the system state must be available in real time. Cascading blackouts can affect national, regional, or continental power systems. Therefore, efficient tool for detection and prevention of disturbances which can cause wide-area blackouts must be developed and deployed. As the blackout risk can be represented as the multiplication of the blackout probability and the cost it produces, effective solutions should be found in order to decrease both factors.

WAMS offers multiple benefits such as stability, reliability, and safety of supply. Also, it has economical effect if the system is operated closer to the grid's stability and capacity limits, which leads to the increased energy transfer.

General trend is to operate power grid close to its stability limits in order to increase power transfer capacity, consequently decreasing safety margins and increasing disturbances and blackout risk. Realistic solution can be found in real-time wide area monitoring, which is based on synchronized phasor measurements (amplitude and phasor angle) of voltages and currents and their frequencies.

WAMS measurement data enable real-time monitoring and it can be used as an early warning system – system which uses a fast diagnostic and gives the operator enough time to make required steps for system stability in order to limit disturbance's range and impact and to prevent power system blackout.

Analyses of the major system blackouts in the last few decades show that the operator mistake was the initial event in about 10% of all the cases. Taking into consideration blackout cost and consequences, even minor reduction in the blackout risk would result in vast savings. WAMPAC has an indirect impact on the system blackout caused by the operator mistake. The operators have the main role in the safe system operation and it can be assumed that they are highly qualified, educated, and monitored in their work and that accidental mistakes happen rarely. Main goal of the quality monitoring system is to supply the operator with the real-time data about the power grid and its state in order to ensure him/her enough time for quality decisions and fast reactions.

Intention is to use WAMPAC and implemented concepts of SIPS to discover and treat real-time disturbances in wide area in order to prevent system blackout and maintain integrity of the whole power system, not just locally or focused on particular element of the grid.

Direct WAMS impact on the system reliability and blackout risk cannot be easily or explicitly expressed, but increased accuracy, additional functions and applications, chronological event registration with time stamps via GPS, as well as future developments in the protection and control field, would significantly improve power grid operation and play main role in decreasing blackout risk.

3.2 Smart transmission grid

The STG represents active infrastructure capable of increased transfer and distribution of electric energy, including considerable amount of energy being from RES. It implies a number of technologies which provide high-quality protection and control, consequently lowering traditionally high safety margins.

Decisions and investments to be made in the next few years will determine development of the power system for decades to follow. Thus, analyzing new concepts as well as implementing them is an important factor in addressing the energy issue of the future. Main intention of many R&D projects is contribution to the development and deployment of the STG, which will comprise elements such as self-restoration feature by means of predicting possible disturbances and returning to normal state by using real-time data, motivation of consumers to actively participate in power system operation, resistance to disturbances caused by human action or natural phenomena, improved integrity, availability and stability of the power system as a whole, providing support for the development of electricity market, and more efficient transmission system regarding market needs and demands alongside with an optimized utilization of all available resources.

Due to many instances of the PTS blackouts caused by extreme loads in deregularized markets, emphasis is placed on the concept of adaptive protection.

3.2.1 Demands and requirements

STG is the future of power transmission. It is a version of PTG capable of satisfying new protection and operation trends of the free and deregulated electricity market. Designing and implementing the concept of STG has come out from the need for developing a system that would eventually reduce the number of large-scale catastrophic electricity supply interruptions and generally improve the safety and reliability of generation, transmission, and distribution of electrical energy in the new market conditions operation of power systems. The STG is a set of technological interdisciplinary solutions, i.e., an infrastructure capable of handling increased transmission and distribution of electrical energy. The concept also comprises the implementation of technologies that enable qualitative monitoring, protection, and operation. Nowadays, it is often pointed out that the aim of STG is to create a system that will contribute to and stimulate global economic growth in respect to sustainable growth and environmental protection.

Decisions and investments concerning PTS which are going to be made in the next few years will determine energy development for centuries to follow. This is all due to the complexity of the EPS. Thus, analyzing new concepts, as well as implementing them, is an important factor in addressing the energy issue of the future.

STG should comprise the following elements:

- self-restoration feature by means of predicting possible disturbances and resetting to normal conditions by using real-time data;
- motivation of consumers to actively participate in grid functioning;
- resistance to disturbances caused by human action or natural phenomena;
- improved availability and stability of electrical energy supply;
- providing support for the development of electricity market;
- more efficient PTG regarding market need and demand alongside with an optimized utilization of all available resources.

STG will largely influence the availability and stability of the entire EPS. Many fields and technologies require research, development, and implementation. Some of the fields are power control, integration of RES, energy storage, direct or indirect load control, relay protection systems, diagnostic and information tools, and data storage.

When trying to find the best solution that can be implemented, challenge lies in the size and dynamic nature of the EPS. Mutually connected systems implement hundreds of thousands of equipment pieces and hundreds of human organizations. The whole EPS is spread over thousands of kilometers with time steps ranging from milliseconds to years.

Intelligent protection and control systems of the PTS include the use of a complex information exchange system in order to prevent or neutralize the spread of wide system disturbances. Basics of protection and control systems are SIPS and SMT of electrical quantities of the power system, which began to be applied in accordance with the rapid development of information and communication technology ICT.

3.3 SIPS – in general

The existing fast relay protection is focused on the individual components of the PTS, such as power lines, transformers, etc. Actions of the individually focused protection often result in the energy delivery blackout, in order to minimize the damage of the equipment. Several cases have shown that the local actions, without having the integrity of the power system as the main priority, can result in the large cascading blackouts [1,2].

The SIPS is a specialized protection scheme which differs to a great extent from the common protection, especially in its main intention. It is designed for integrity protection, failure prevention, and mitigation of disturbance consequences in the whole PTS. This requires specialized circuitry and software solutions, which cannot be obtained from the market as a finished product, but have to be developed and tested in real time. Input parameters for establishing the SIPS in the PTS could be (and in this case will be) synchronized measurements of voltage and current phasors, obtained from the PMUs, as a globally acknowledged STG technology. In contrary, the local protection schemes are oriented towards local measurements of RMS values, and thus do not ensure the power system integrity in case of a fault.

The realization requires that the SIPS is added to the existing WAMPAC, based on synchronized measurements [3].

The SIPS architecture can be distributed or centralized, based on events, parameters, response, or combination. Also, there are two basic solutions: the flat (measuring and operating elements are in the same location) and hierarchical architecture (local measurements are sent to multiple control locations).

Although the SIPS can help to increase power transfer limits, its primary task is to reduce the intensity and frequency of blackouts and to improve security of the power system. Rapid development of technologies for the PTS protection and control, such as communications, controllers and PMUs, and analytical techniques for applications, has enabled the implementation of the SIPS concept, which can encompass several applications [4].

3.4 Wide area disturbances

When a major disturbance occurs, protection and control actions are required to stop the power system degradation, restore the system to a normal state, and minimize the impact of the disturbance [5]. Control center operators must deal with a very complex situation and rely on heuristic solutions and policies.

Generally disturbance involves a combination of phenomena listed below:

- Cascading line tripping by dynamic line loading – often lends to unsuccessful line restoration attempts;
- Cascading equipment tripping by overexcitation;
- Loss of synchronism due to angle instability;
- Oscillatory instability causing self-sustaining interarea-oscillations;
- Exceeding allowed frequency range (over and underfrequency) due to imbalance in active power between generation and load;
- Voltage instability/collapse.

Using the SMT and wide area monitoring, the propagation of disturbances throughout the system will become more easily manageable. This will open up new opportunities for designing novel early warning systems and closed loop control applications, which will contribute to the prevention of cascading and catastrophic blackouts The disturbance in the power system usually develops gradually, but the time scale in which the dynamics of the disturbances affects the underlying power network may be from several minutes to milliseconds, depending on the type of the disruptive event. In all cases, the sequence of actions needed for effective protection and/or emergency control (between which the boundary may sometimes be overlapping) consists of the following elements:

- Identification and prediction (often needed to initiate a real-time action);
- Classification and location of the disturbance(s);
- Decisions and actions to be undertaken to arrest the degradation of the system state, assure a safe return to the secure state, and return to economical and environmentally sound operating state.

Coordination of various actions must be done (this step could sometimes present a challenge); corrections should be planned should the result of automated actions result in the nonsecure, noneconomically optimal, or otherwise undesirable state.

The knowledge of the complete state of the power system, represented by several network parameters, requires real-time state measurements as an input. As it is often unavailable, partially due to legacy technology and associated complexity, a partial set of necessary inputs is often used to make decisions related to system protection thresholds, trends, patterns, and sudden changes of these parameters to provide key information to detect an emergency. Among the parameters are:

- Active (or reactive) power flows in the network;
- Reactive power reserves at major power-generating plant;
- Voltage magnitudes in selected locations (effective for detecting equipment overloads, voltage collapse conditions, etc.);
- Phase angle differences between locations of interest (often ends of major transmission links, used for detection of out-of-step conditions and development of cascading events);
- Driving point impedances from certain locations (used, for example, in designing out-of-step tripping or blocking schemes);
- Resistances and their rates of change (used for speed-up or out-of-step detection);
- Frequency and rate of change of frequency;
- System spinning and cold reserves (for underfrequency load shedding schemes, system separation, and controlled islanding schemes); system load, meteorological and seasonal factors, as well as status of relays and breakers in the network are also used in formulation and design of the system protection schemes for wide area event horizon.

The protective and emergency control actions are selected from a number of possibilities, which includes out-of-step relaying, load shedding, controlled power system separation, generation dropping, fault clearing, fast valving, dynamic braking, generator voltage control, capacitor/reactor switching and static VAR compensation, load control, supervision and control of key protection systems, voltage reduction, phase shifting, tie line rescheduling, reserve increasing, generation shifting, high-voltage direct current (HVDC) power modulation, etc. [6]. As a disturbance progresses and the state of the system degrades, less desirable actions may become necessary. All of the available means of protection and control are suitable during transition to or within an "in extremis" situation. Preventive (less intrusive) measures are usually the only measures appropriate for an alert state. The set of actions is implemented in the emergency procedures for the power system. Every system has its own emergency control practices and operating procedures dependent on the different operating conditions, characteristics of the system, and engineering experience. The operating procedures for every system are unique (heuristic procedures are often used).

In an environment where the protected area is large, it would be very hard to design a protective or emergency control scheme based on fixed parameter settings. Adaptive approach is preferred in such circumstances, possessing ability to adjust to changing conditions. Relays which participate in wide area disturbance protection and control should preferably be adaptive. Relay system design should satisfy minimum requirements of having the ability to communicate with the outside world. The communication links must be secure, and the possibility of their failure must be allowed for in the design of the adaptive relays. The system measurements used in the relays must be related to the parameters which help observe the disturbance propagation. Such measurements must provide information about changing system conditions so that they will be useful in the management of the disturbance.

Besides monitoring of phasors of voltage and branch currents, the PMUs can track local frequencies at different locations, providing useful data for estimation and location of disturbances for further control actions. These may be load shedding or increment/reduction of power generation from/to other areas.

3.5 SIPS architecture

Several papers published by the Institute of Electrical and Electronics Engineers (IEEE) [6] define System Integrity Protection Scheme (SIPS):

"The SIPS encompasses special protection system (SPS), remedial action schemes (RAS), as well as other system integrity schemes, such as underfrequency (UF), undervoltage (UV), out of step (OOS), etc."

SIPS classifications have been defined through a collective global industry effort by members of the IEEE and International Council on Large Electric Systems (CIGRE) [6]:

Local (Distribution system) – SIPS equipment is usually simple, with a dedicated function. All sensing, decision-making and control devices are typically located within one distribution substation. Operation of this type of SIPS generally affects only a very limited portion of the distribution system such as a radial feeder or small network.

Local (Transmission system) – All sensing, decision-making and control devices are typically located within one transmission substation. Operation of this type of SIPS generally affects only a single small power company, or portion of a larger utility, with limited impact on neighboring interconnected systems. This category includes SIPS with impact on generating facilities.

Subsystem – The operation of this type of SIPS has a significant impact on a large geographic area consisting of more than one utility, transmission system owner, or generating facility. SIPS of this type are more complex, involving sensing of multiple power system parameters and states. Information can be collected both locally and from remote locations. Decision-making and logic functions are typically performed at one location. Telecommunications facilities are generally needed both to collect information and to initiate remote corrective actions.

System wide – SIPS of this type are the most complex and involve multiple levels of arming and decision-making and communications. These types of schemes collect local and telemetry data from multiple locations and can initiate multilevel corrective actions consistent with real-time power system requirements. These schemes typically have multilevel logic for different types and layers of power system contingencies or outage scenarios.

3.5.1 Design

The design of different SIPS depends on the type of scheme and the specific conditions under which it is applied. However, the design process is similar due to the fact that it needs to consider the requirements of the application and based on them to define how they will be satisfied by the design of the scheme.

The design of SIPS should address all standard requirements for protection terminals [7, 8]. The terminal is connected to the substation control system. For time tagging applications, a GPS-based synchronization function is needed. The system protection terminal possesses a high-speed communication interface to transfer data between the terminal databases, which contain all updated measurements and binary signals recorded in that specific substation. The conventional substation control system is used for the input and output interfaces with the power system. The decision-making logic contains all the algorithms and configured logic necessary to derive appropriate output control signals, such as circuit-breaker trip, AVR-boosting, and tap-changer action, to be performed in that substation. The input data to the decision-making logic is taken from the continuously monitored data, stored in the database. A low-speed communication interface for supervisory control and data acquisition (SCADA) communication and operator interface should also be available as an enhancement for the SCADA state estimator. Actions ordered from Energy Management System (SCADA/EMS) functions, such as optimal power flow, emergency load control, could be activated via the system protection terminal. The power system operator should also have access to the terminal, for supervision, maintenance, update, parameter setting, change of setting groups, disturbance recorder data collection, etc. For local schemes, where monitoring and decision stations are within close proximity, there may still be a need for use of high-speed communication (Table 3.1).

Figure 3.1 describes fundamental SIPS architecture as part of WAMPAC system.

Table 3.1 Actors which are referenced in Figure 3.1

Number	Actor name	Actor type (person, device, system, ...)	Actor description
1	Field devices (PMU)	Device	PMUs are devices capable of measuring voltage and current sinusoidal waveforms on transmission lines, and transmitting the data to the utility for monitoring and control purposes. The data consists of voltage and current measurements (real and imaginary) and frequency. The data is accurately time-stamped to IEEE standards. Also mitigating devices that acts to field equipment (e.g., circuit breaker) upon command from PDC based on SIPS application result.
2	Data alignment	System	Raw synchrophasor data are aligned, archived and distributed to other phasor measurement network parties by PDC.
3	Database 1	System	Database 1 is a common data repository for raw synchrophasor data.
4	Data processor	System	This is a system that is used to manage, process, and respond to dynamic changes in fast-moving streaming phasor data. More specifically, it can process raw synchrophasor data.
5	SIPS applications	Process	The SIPS applications are set of software modules within PDC system that runs certain calculations in order to process raw synchrophasor data.
6	Data archival system	System	The data archival system is a system with set of functions that enables archiving and retrieval of rough PMU measurements, application results, and alarms.
7	Database 2	System	Database 2 is a real-time data repository for all raw synchrophasor data and nonoperational data such as application results and alarms.
8	Visualizations (Displays)	Display	Main phasor measurement system display that visualizes raw synchrophasor data, applications data alarms, and other data defined by user. Main display gives overall real-time monitoring visibility of the bulk power system.
9	Clients (Engineering, operations)	Clients	Operations – clients who "operate" the power system using phasor measurement system application and EMS, SCADA applications. Engineering – clients whose knowledge about the power system advise operators (usually upon request) before and during their execution of complex or unusual procedures.
10	Energy management system – EMS	System	The EMS is a system of tools used by system operators to monitor, control, and optimize the performance of the transmission system. The monitor and control functions are performed through the SCADA network. Optimization is performed through various EMS applications.
11	Power system database	System	Power system database is defined by topology (connectivity among power system components such as generators, power transformers, transmission lines, loads) and power system parameters (line, transformer, generator parameters, etc.). Power system topology changes during certain period of time and database should be changed accordingly.
12	Clients	Various	

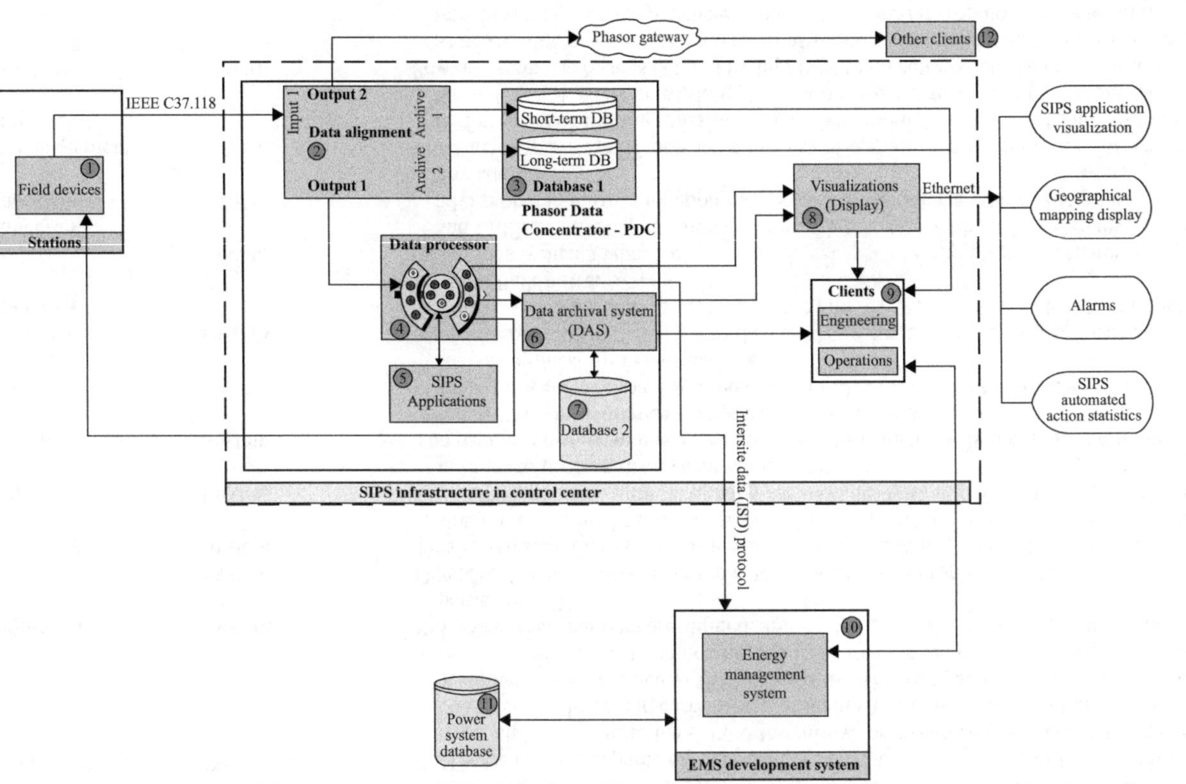

Figure 3.1 SIPS architecture

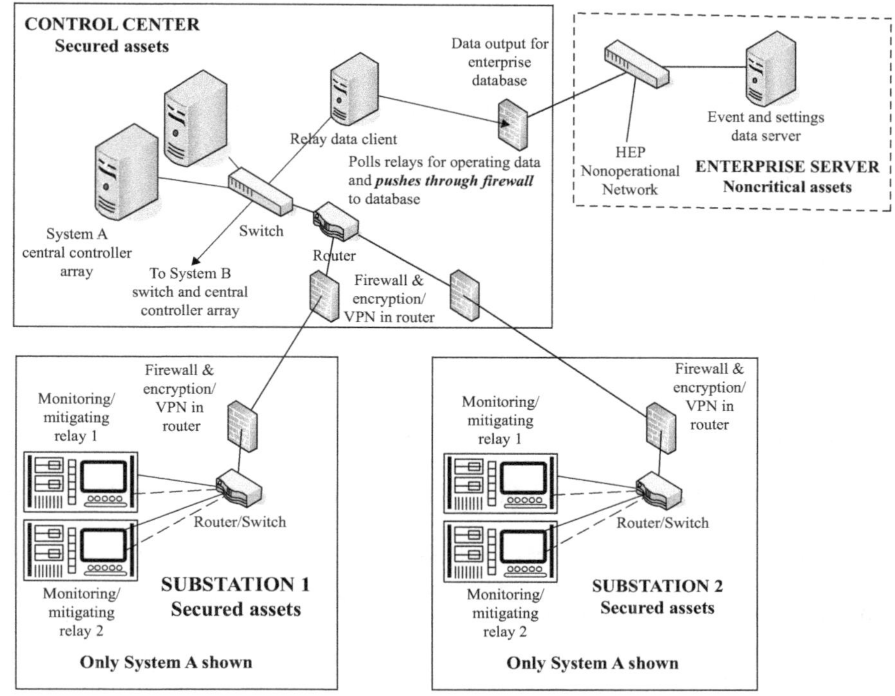

Figure 3.2 Redundant SIPS architecture

Since SIPS is expected to perform mission critical functions in spite of equipment maintenance or failures, it has to be realized in a dual redundant configuration [9]. The objective of dual redundancy is for the SIPS to continue to perform its required functions, with no significant degradation of performance and no loss of data, following the failure of any single SIPS component (central controller – PDC, field device – relay, or communication circuit). For some components such as controllers or communications networking connections within a facility, additional redundancy may be designed into each of the two SIPS systems. For example, the central controllers (PDC) in each of the separate systems are sometimes configured as dual redundant or triple redundant to achieve availability and security approaching that of the protective relays in the substations.

The dual redundant system should include the following major components (Figure 3.2):

- Dual redundant monitoring relays, System A and System B, at each substation in the SIPS plan. These relays supply real-time information about transmission system flows, breaker states and operations, and contingencies (fault protection relay operations). Where a large number of values are to be gathered, multiple relays may be required for each of System A and System B.

- Dual redundant-programmable central controllers – PDCs that process information received from the SIPS line monitoring relays to determine if any mitigating control actions are needed – System A and System B. Each central controller – PDC will be designed to handle the SIPS analytics, data historian, external system interfaces, and input conditioning completely independent of the other controller.
- Dual-redundant mitigation relays – System A relays and System B relays that execute the control actions requested by the SIPS central controllers – PDCs at substations in the SIPS plan. These can be the same relays that perform the monitoring.
- Redundant communication links that carry the measurements, status indications, and control commands between the SIPS substations and the SIPS central controllers.

The SIPS communication network has to support a wide ETHERNET area (WAN) that will transmit Layer 2 multicast IEC 61850 GOOSE messages for high-speed measurement and control, and a third-level IP-based network with data for system monitoring and control.

The network will be completely redundant and diversified, with two separate paths from each substation to the transmission system operator control center, without a single point of failure that could negatively influence the multiple redundant components or paths. The realization of all SPS components has to be constantly monitored and failure has to be alarmed. The system has to be immune to "broadcast storms" and malfunctions that are not alarmed, like controller blockade in code loop. All equipment has to support NERC CIP cyber security standards.

3.5.2 Multipurpose open SIPS architecture

Development of SIPS for processing, archiving, and visualization of phasor data started with the idea of extending archiving capabilities in a way which would not require any changes in existing system [10]. This was realized with custom historical data server which acquires data from existing system through OPC–HDA interface and stores data into external MS SQL database which can hold a much larger amount of historical data than existing historical database and is practically limited by available disk size. Additionally, data archiving is event driven with customizable parameters for each measurement or calculation result. In order to further minimize number of unimportant events and thus the size of historical database, hysteresis range parameter is applied on input data which prevents rapid event generation in case of oscillation of data values around threshold value. Events triggered in the process of analyzing input historical data are also stored in database which simplifies later identification of important points in historical database.

Next step in the development of SIPS was to extract real-time data from existing system and potentially other data sources, do calculations, make mitigation action, and visualize data to the operator.

The main and the most difficult task to be accomplished was the modeling of a multipurpose and open SIPS architecture which would be able to communicate and

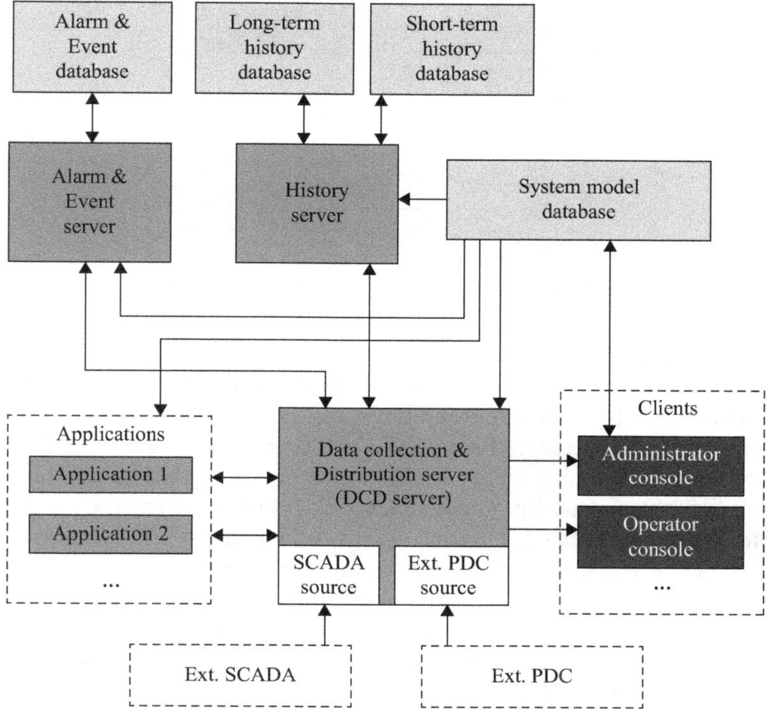

Figure 3.3 Multipurpose SIPS architecture block diagram

collect data from different sources such as the already existing PDC, virtual PMUs, SCADA, and synchrophasor vector processor [11]. The system should manage data from all of these sources, filter data if required, provide real-time monitoring, create record of outages and errors, record, and store data [12, 13]. Also, the human interface is needed to be adjustable for different user profile requirements. The system architecture block diagram is shown in Figure 3.3.

System architecture is based on the concept of universal data representation and loosely coupled components to achieve maximum flexibility in customizing system for different user requirements without changing the core of the system [14].

System is built around the main information data store (system model database) which is used by every system component and which contains information such as classes of data, network model, visual model, application parameters [15, 16]. Data store is based on universal data representation in a form which bridges object oriented higher-level data representation with relational database and essentially represents a form of object-relational database. System model database is physically implemented as MS SQL database with stored procedures used for reading and writing data. Data representation model is shown in Figure 3.4. Data is represented through four entity types: resource, link, property,

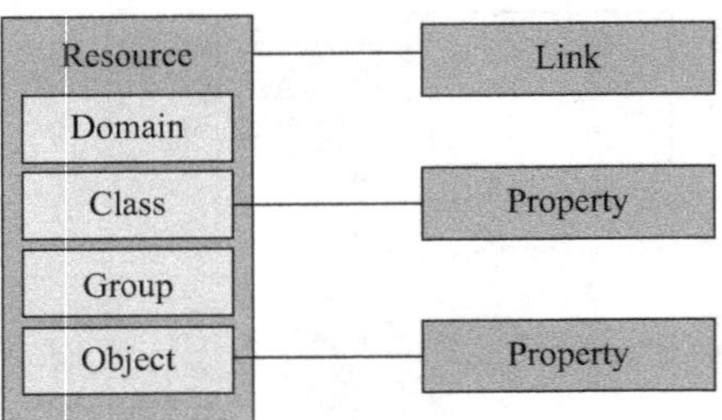

Figure 3.4 Data representation model

and property value. Resource is a general entity type which represents any type of abstract or real object in model. Resource is divided into four types: domain, class, group, and object. Domain and group resources are used for hierarchical grouping of classes and objects, respectively. Class and property resources together define template for objects which represent real-world objects with property values containing actual object data. Resources can be linked together with link entity to form relations between domains, classes, groups, and objects. Link entity can link resources in two ways: vertical (parent–child) relation and horizontal (connection) relation.

The advantage of this concept is that the structure of power system model as well as other types of data structures is not predefined and can be changed by a user through graphical user interface to fulfill the needs of particular application. For example, if the user wants to add new custom calculation application to the system which uses some new power system parameters, he/she will simply expand the model with these new parameters through graphical user interface. Another advantage is that extension of the system with new calculation applications, visualizations, and data sources does not require any changes to the core of the system. This advantage has already been realized in the process of system development where core system components almost did not change since the start of development.

Central part of the system for handling real-time data is data collection and distribution server (DCD Server) which routes data between different data sources such as external PDC or SCADA and external calculation applications and clients. DCD server is based on publish/subscribe mechanism which allows event based data exchange between different system components. There are two types of components which connect to DCD server: data source and external application. Data sources are special plug-ins which implement required interface for DCD server on one side and specific communication protocol on the other side to enable communication with external data sources. External applications are administrator and operator console for system configuration and data visualization, history server for data archiving, alarm and event server for alarm and event processing and different calculation applications.

Real-time data processing and calculation is performed within calculation applications which run in the background as operating system service. They connect to the system model database to get power system network model, parameters, and other information and DCD server which provides real-time data from external sources and other components and allows calculated data to be published and available to other system components. Applications can connect to system model database and DCD server directly or through special adapter libraries which provide customized data access and formats required by particular application. For example, most calculation applications need simple network model with just two object types: node and line. In such a case it would be justified to use adapter library which would act as a layer between more abstract data representation in model database and concrete and simple objects required by calculation application.

Integration of all system components shown schematically in Figure 3.3 is archived by using technology of distributed objects which communicate over Transmission Control Protocol/Internet Protocol (TCP/IP). This automatically enables powerful horizontal scaling of the system if more processing power is needed. For example, each calculation can be built as separate service and placed on different servers without changes to the system configuration except making sure that the correct IP address of the DCD server is entered in each calculation service configuration file.

There are two types of client applications with graphical user interface: administrator console and operator console made for corresponding two types of users.

Administrator console allows system administrators to configure every system component and modify data and visual model. It currently allows editing model structure, the actual data model, visual model, visual symbols, and custom functions. Model structure editor is used for editing domains, classes, and properties of previously described data model. It is essentially a template editor which is used to define all structures which will be used to represent actual objects and data. Editor allows creation of domain and class hierarchy with definition of properties for each class and possibility for one class to inherit properties of each parent class. For each class property, various attributes can be defined like, for example, type of value which can be real number, text, image, file, or something else. Another part of administrator console is a model editor which uses previously defined classes as templates for actual objects which represent real-world power system objects or other types of objects which are used internally like diagram, function, etc. Administrator console with model editor is shown in Figure 3.5.

Model editor of administrators console allows creation of hierarchical tree of objects and object groups which is represented by parent–child relation in model database. For each object it is possible to specify any number of classes, parent objects, groups, and connected objects. Actual object data is stored in form of property values which correspond to properties defined in classes specified for particular object. In Figure 3.5, editing of one transmission line object is shown. In the middle of the screen, a number of property values defined by transmission line class are visible. They represent transmission line parameters while measurements and calculated values are contained within separate child objects. It should be noted

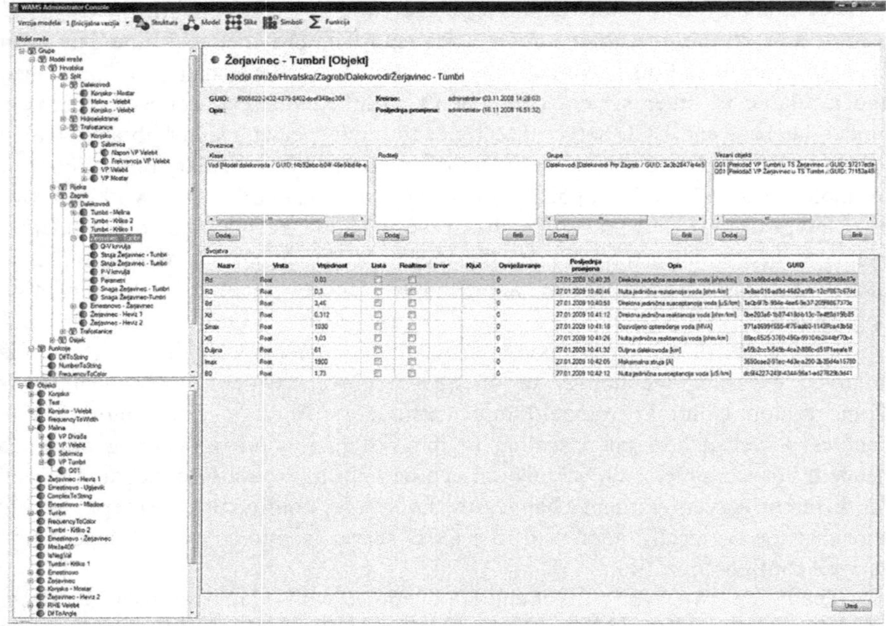

Figure 3.5 Model editor of administrator console

that the value of any real-time property is not stored in model database. It is only identified by its globally unique identifier, used in data exchange between DCD server and various applications and in the end eventually stored in historical database.

Visual model editor is another part of administrator console shown in Figure 3.6, which allows editing of visio-style diagrams which can represent any type of vector graphics. Diagram shapes can be drawn using various basic tools or made from predefined or custom shapes which can be edited through symbol editor. The most important feature of the diagram is that it can be made dynamic by binding real-time data to shapes. This binding is achieved between shape property and custom function with real-time values as parameters which returns value of shape property. Since each shape has plenty of properties, powerful dynamic visual diagrams can be created in this way. For example, a user can build voltage angle difference gauge with dynamic arrow which will change its angle proportionally to the voltage angle difference real-time data. Also, the gauge can change color if angle difference exceeds predefined limit.

The functions used currently for visual elements can be written in C# language and are edited in separate part of administrator console. It is a rich-text editor with syntax coloring and common text manipulation functions. Functions can take any number of real-time arguments and parameters to calculate resulting value. Since

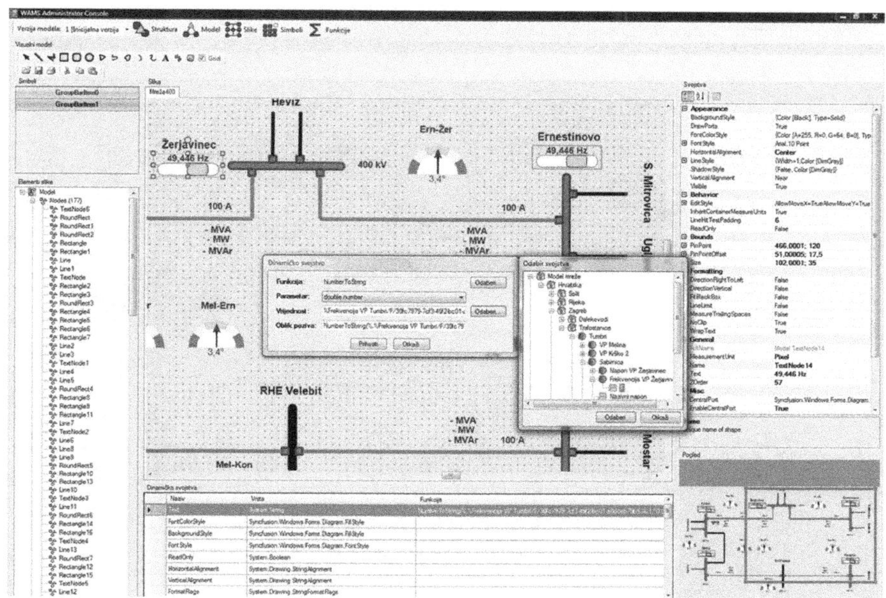

Figure 3.6 Visual editor of administrator console

the function language is C# and functions are compiled with standard .NET C# compiler, there is a great flexibility in writing function code because of the .NET framework in the foundation.

Operator console is a graphical user interface for system operator and analyst designed with simplicity and interface customization in mind and is used for graphical and textual real time and historical data presentation. Operator console allows users to create personalized views of data by using customized data presentation modules.

Data presentation modules are separate plug-ins for operator console which implement required interface and are completely independent applications for data presentation. Each module can be customized and customization data can be saved to disk which allows building different looking versions of the same presentation module. Modules are displayed in separate windows while the whole workspace is multiwindow with an option to save positions and types of displayed data presentation modules for later switching between views. Operator console with some opened data presentation modules is shown in Figure 3.7.

History server is another important part of the system which enables power system performance and post disturbance analyses by recording measured and calculated data from different data sources and applications to SQL database. Currently, it is planned to extend current history server which independently archives data from an existing system and to integrate it with new system. It will have more advanced features like complex rules with multiple parameters written as C# functions to enable event-based data recording for more efficient data

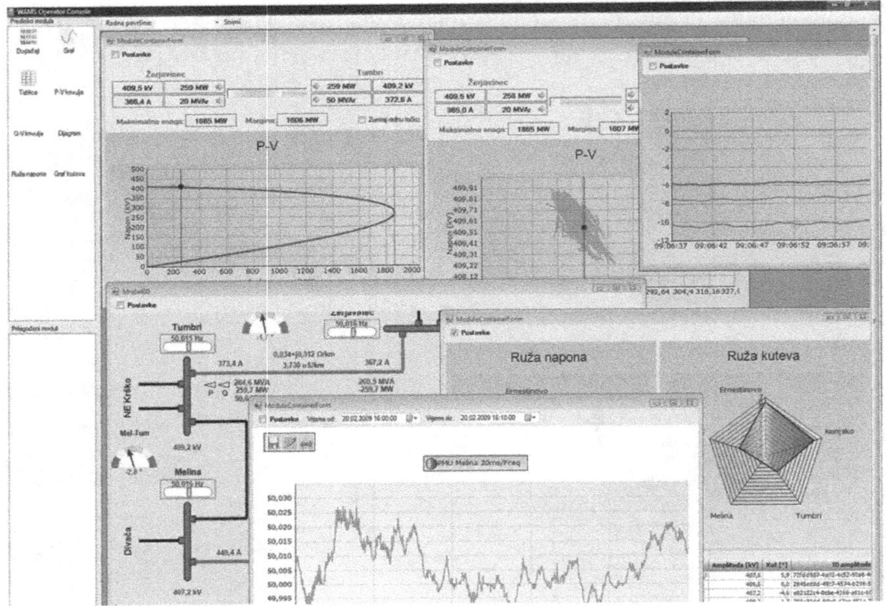

Figure 3.7 Operator console

storage. Also, it is planned to introduce new data source for DCD server which will connect to history server and distribute historical data to subscribed clients as real-time data in the same way as any other real-time data source. This will be the extension of currently built data source which allows replay of historical data from CSV files.

History server reads and writes data into two SQL databases. Short-term history database is the place where data is stored first and it is a relatively small database with maximum history of one day. Because of its smaller size it is much faster to use for frequent history operations like real-time data insertion. Data from short-term history database is periodically transferred to long-term history database which is much larger and thus slower but it is not used so frequently as short-term database. History databases are cyclic with configurable data retention interval which means that after defined period of time the oldest data will be automatically deleted.

Last part of the system which is currently under development is advanced alarm and event system which will enable real-time recognition of important events by using previously described custom functions. This way it will be possible to recognize event which is triggered by particular combination of various real-time measurements and calculations. Also, it will be possible to define appropriate alarm levels for such events.

All events and alarms will be automatically stored in separate SQL database which will allow later identification of important historical events and also simplified navigation through historical data.

3.5.3 SIPS components

For a single-purpose SIPS, it is most efficient to place the analog threshold decision platform in the substation where analog measurements are to be taken, since these measurements would require more complex communications channels to transport values to a remote site at high speed than binary points require. For example, a relay at a line terminal substation will detect when line loading exceeds a target value developed from an a priori system study. The arming indication that the target load was exceeded is transmitted as a simple binary point to a remote site where a contingency such as a line or generator outage might then be detected. At the remote site, the logical combination of the arming indication binary state and the line or generator outage binary state quickly lead to load shedding at a third remote load center. No analog information needs to be transmitted at high speed. The SIPS can react in well under 100 ms.

However, as soon as the SIPS requires comparison or addition of analog values from multiple separated sites, or requires flexibility of evaluation strategy for analog values, then these must be transmitted among sites. Flexibility requires that all values be sent to one or more redundant central controllers, each placed at a communications nexus where cost and time delay of the communications is minimized.

Even when the design focuses on highest practical speed for transmission of all critical information, binary points can always be communicated faster and/or updated more frequently than analog values. Therefore, the design of the logic or analytic for an SIPS often uses analog values for *arming – taking no immediate action, but setting up the SIPS to operate if a contingency such as a breaker trip occurs*. Then, the reaction to the contingency, communicated as a binary point, is very fast. In some conventional SIPS designs where tracking of analog measurements does not require fast detection of sudden value changes, SIPS arming measurements are performed by the SCADA or EMS and communicated to an SIPS relay via an RTU control point. The communications channels for critical binary values can then be high-speed digital communication channels similar to those used for pilot or unit protection of transmission lines.

This strategy cannot always be applied – if the contingency is overload of a system element without one specific arming cause or circumstance, the analog values surrounding that element must be processed in real time to make a decision on the need to act.

3.5.3.1 Monitoring and mitigation devices (relays)

Protective relays are the sensing and mitigation devices in substations are protective relays. Features of modern microprocessor relays important to this application are:

- Scanning cycles for binary inputs of 2 to 4 ms plus debounce delays;
- Repetitive updating of analog measurements as often as once or twice per power cycle (synchrophasors in IEEE C37.118 format) or 2 to 4 times per second (metered values in SCADA protocol or IEC 61850 analog GOOSE messages);

- Data communications in multiple formats for direct connection to serial or Ethernet channels;
- Control outputs for breaker tripping that operate from data communications or from local decisions;
- Programmable logic that can be configured for threshold detection, trip output voting or security, and monitoring of the operating integrity of the relays and communications paths;
- Environmental hardening for reliable substation installation;
- Familiarity to protection and control technicians.

Monitoring relays installed at measurement substations around the transmission network provide line or breaker status reports and system measurements to the central controllers.

Mitigation relays serve as control devices acting on command from the central controllers.

SIPS relays should be separate from those used for fault protection:

- Most practical for maintenance and testing of either the SIPS or fault relays without endangering of too many functions that are critical to system operation;
- The risk of a misoperation from technician maintenance error is reduced;
- The SPS can be installed and commissioned as a separate activity.

At each SIPS location, the installation should feature dual redundant relays, installed so that there is no credible single point of failure that can completely disable SIPS operation.

3.5.3.2 Central controllers

Redundant *central controllers* exchange information and control actions with the monitoring and mitigation relays via a reliable and secure communication network. In general, the monitoring information may come from different substations than those where mitigating actions are taken. The central controllers also provide interfaces for personnel:

- Operators who monitor the SIPS performance and enable or disable operation of various protective functions (analytics);
- Maintenance personnel who maintain or restore SIPS component integrity;
- Programmers from planning department who configure the logic and measurement analytics based on a priori planning studies.

Each central controller is a programmed logic platform with connection to relays and operators via communications paths as specified in other sections. Several types of devices can be considered:

- Protective relays with communications ports and logic programming capability, up to the logic processing capability limit;
- Programmable logic controllers (PLCs) with added communications interfaces;

- Personal computer (PC) or related server-type computing platforms, with added operating systems, logic programming applications, data communications interfaces and processing, and data storage and management applications.

3.5.3.3 Data communications network

A communications network with redundant paths is required to connect all the monitoring and mitigating relays to the central controllers. While there are broad choices for a user building a new installation, optical fiber communications network with at least two paths to all substations is identified for the SIPS activities. The network must be equipped with channel interfaces for the selected communications protocols.

3.6 SIPS data archival system

Data archival system (DAS) or historians are optimized to effectively handle large volumes of time-stamped measurement data typically used to save and retrieve phasor data but should be able to deal with SIPS application results data, event data, and alarms data.

The volume of streaming phasor data can make storage requirements add up quickly. For ease of retrieval and use, data is stored as time-value pairs either in a data historian or in a relational database. Not counting data structure overheads, the minimum possible size storage is 10 bytes per time-value pair (4 bytes for time, 4 for data, and 2 for flags) within a historian. Thus a PDC that collects data from 100 PMUs. Of the 20 measurements each at 30 samples per second, will require a little over 50 GB/day or 1.5 TB/month of storage within a historian system. Long-term database should be able to store synchronized measurements up to 5 years which dramatically expand needed data storage infrastructure. It has to be outlined that calculation results from applications that use synchrophasor measurements should be also archived. Assuming 100 PMU and archiving different kind of data, described above, in five years database should be able to store more than 500 TB of synchronized data. Figure 3.8 shows a structured diagram of DAS databases based on the time duration of the stored data.

Primary data storage is directly accessible to the CPU. Any data actively operated on is also stored there in uniform manner or dynamic random access memory (DRAM). DRAM capacity can be developed up to order of 10 GB and could be combined in slots. Since 2006, solid state drives (based on flash memory) with capacities exceeding 64 GB and performance far exceeding traditional disks have become available. This development has started to blur the definition between traditional random access memory and "disks", dramatically reducing the difference in performance.

Up-to-date storage technology capacity for a hard disk drive (HDD), secondary data storage, is controlled by the size of the drive, the "areal density" of the drive and the number of platters.

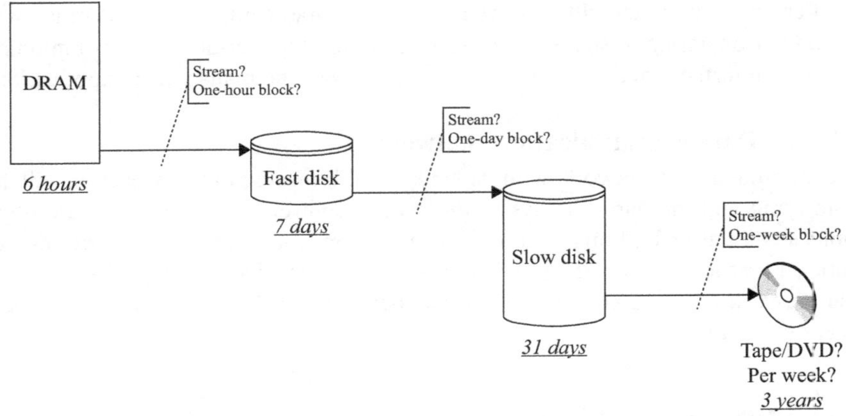

Figure 3.8 Structured diagram of DAS databases based on the time duration of the stored data

Off-line storage is a computer data storage on a medium or a device that is not under the control of a processing unit. The medium is recorded, usually in a secondary or tertiary storage device, and then physically removed or disconnected. In modern PCs, most secondary and tertiary storage media are also used for off-line storage. Optical discs and flash memory devices are most popular, and to much lesser extent removable HDDs. In enterprise uses, magnetic tape is predominant. Older examples are floppy disks, zip disks, or punched cards.

From computational technology perspective, maximum capacity is not the primary goal, but speed of data archiving and retrieval. As a result the drives are often offered in capacities that are relatively low in relation to their cost and potential.

SIPS system archival requirements and data storage technology constraints forced DASs to use various data storage techniques.

The phasor community is beginning to develop standard formats for phasor data, or changing existing formats to accommodate phasor information. For example, the widely used COMTRADE format is being tuned so that phasor data can be effectively stored in this format and used with tools that are COMTRADE compatible.

Currently there are no formal data retention standards for phasor data. Current phasor system managers are keeping full resolution phasor data archived for at least two years (preferably 5 years) to facilitate disturbance investigation and research, with data pertaining to disturbances retained longer.

3.6.1 Real-time database

DAS as part of overall SIPS architecture is based on the real-time data collection, archiving, and distribution engine or more specifically on real-time database models.

Traditionally, real-time systems manage their data in application-dependent structures. As real-time systems evolve, their applications become more complex and require access to more data. It thus becomes necessary to manage the data in a systematic and organized fashion. Database management systems provide tools for such organization, so in recent years there has been interest in "merging" database and real-time technology. The resulting integrated system, which provides database operations with real-time constraints, is generally called a real-time database system (RTDBS).

Similar to a conventional database system, an RTDBS functions as a repository of data, provides efficient storage, and performs retrieval and manipulation of information. However, as a part of a real-time system, whose tasks are associated with time constraints, an RTDBS has the added burden of ensuring some degree of condence in meeting the system's timing requirements.

According to real-time database philosophy, DAS shall bring all relevant data from disparate sources (PMU data, application results data, events data, alarms data, external sources), into a single system, and shall have ability to give access and delivers it to individual modules of SIPS system based on their roles in a uniform and consistent manner. DAS shall handle high-speed, real-time, data collection and can collect data at subsecond speeds by leveraging the latest 64-bit technology, multiprocessor/multicore hardware, and Windows server operating systems. Data should be instantly stored in the data acquisition and distribution server and made available to users in real time. In addition, the data acquisition and distribution server should be able to make available decades of data and can store data for millions of points.

DAS shall be able to optimize the data storage, to use the least amount of computing resources possible while still providing needed fidelity and allows retrieval of any data, no matter how old, quickly and accurately.

DAS real-time operation means that DAS shall ensure both continuous storage of the data and continuous availability to all the data for all users.

3.6.2 DAS architecture

DAS architecture shall satisfy requirements on the time duration of stored SIPS data:

- Real-time memory resident database;
- Short-term database;
- Long-term database

and requirements on various type of data:

- Synchrophasor database;
- Application results database;
- Event database;
- Alarm database;
- SIPS system event database; and
- Power system network model database

The functionality of the SIPS system will be distributed by components interacting with each other. The system should consist of a service for collecting, distributing, and processing data, a database for storing system parameters, network model and the measured and calculated data, services for clients with a graphical interface for data display. The system must support the collection of real-time data from external data sources, and must also enable and transmit data in real time to external data recipients.

Specific requirements on DAS

- The architecture of the SIPS system must be service oriented, which means that the functionality of the system must be separated into components that provide certain services through a clearly defined interface.
- Communication between the system components must be based on events (event-driven architecture);
- Database should be run over same platform as the whole SIPS system (e.g., MS SQL server, .NET platform);
- It is necessary to leave the option to the user to set time interval for data storage in databases.
- Upgrading the system which aims to expand support for new communication protocols or new graphical representation must be made in the form of plug-in modules to be easily installed in an existing system to add a desired functionality.
- System components for communication shall use a standard protocols such as . NET Windows Communication Foundation.
- Applications must have all the server names, IP addresses, TCP/User Datagram Protocol ports and other communication parameters that are used to communicate with other applications and components stored in the configuration file.
- Applications must write all errors, warnings, and important information (e.g., start, stop components) as well as security events in the standard Windows Event Log. Additionally, the user must get brief information about the error or event bit in the system over the standard graphical dialogs.

3.7 SIPS applications

The types of SIPS applications may vary based on the topology of the power grid. There may also be different views on the acceptability of the type of the application. For example, use of SIPS for generation shedding to balance grid performance may be viewed as unacceptable for certain levels of contingency in one network but a common practice in another interconnected grid. Consider power systems with limited transmission corridors where building a redundant and diverse interconnection outlet for a generating facility may not be physically practical or economically feasible to address variety of technically possible outlet outages. In such conditions, the generator owner may accept a certain level of risk so long as it can be demonstrated that such SIPS does not result in an unacceptable level of security for to other parts of the grid.

The typical applications are designed to keep a system or subsystem in parallel with the remaining parts of the system in case of the loss of a major supply to the affected power system area. Such major supply deficiency may be caused by:

- Loss of significant amount of generation;
- Loss of important transmission lines or interconnections;
- Overloading of transmission lines or power transformers.

The mitigation measure to maintain grid integrity is described in a document under development by a collaborative effort of IEEE, CIGRE, and EPRI [6]. Below is a summary listing of the types:

- Generator rejection;
- Load rejection;
- Under-frequency load shedding;
- Undervoltage load shedding;
- Adaptive load mitigation;
- Out-of-step tripping;
- Voltage instability advance warning scheme;
- Angular stability advance warning scheme;
- Overload mitigation;
- Congestion mitigation;
- System separation;
- Shunt capacitor switching;
- Tap-changer control;
- SVC/STATCOM control;
- Turbine valve control;
- HVDC controls;
- Power system stabilizer control;
- Discrete excitation;
- Dynamic breaking;
- Generator runback;
- Bypassing series capacitor;
- Black-start or gas-turbine start-up;
- AGC actions;
- Busbar splitting.

3.8 Data protocols

Control speeds must approach those of protective relaying systems. Legacy SCADA or industrial control protocols typically have scan times on the order of fractions of a second to several seconds, and will limit application options for a distributed high-speed SIPS.

For high-speed transmission of only binary information – critical status changes or trip commands – existing transfer tripping channel types could suffice, although this is a fallback plan only if providing Ethernet becomes a problem. If voltages and currents used for arming or a priori determinations do not require high speed, a mixture of protection and SCADA protocol traffic may serve some SIPS applications.

For gathering binary states within milliseconds plus analog values at a reliable rate of many readings per second, the SIPS should use the following protocols:

- IEC 61850-8-1 GOOSE messaging publishing binary status and control point information as layer 2 multicast Ethernet packets – available with many manufacturers' protective relays and some client devices. First response by a substation relay to a binary state change is transmitted in as little as 2 ms, with repeated retransmission. For transmission to the control center, channel delay of 7 to 20 ms will be added depending on equipment, assigned bandwidth, bridging delays, and cyber security of the messaging (e.g., VPN between routers).
- IEC 61850-8-1 GOOSE messages containing analog metered values. Some relay products offer synchrophasor values for transmission in lieu of conventional metered values. Transmitted analog values are updated at a rate of 2 to 10 times per second.
- IEEE C37.118 Synchrophasor streams from PMUs and some vendors' relays – analog values with precision timing/phase position, updated up to 50 times per second (one value per power cycle). Selected new products can stream every half cycle – 100 values per second. C37.118 synchrophasors require PDC and/or phasor data client software products at control center. This is the fastest mechanism for gathering analog values available today, although it requires more intense central controller processing and may thus have a higher cost.

3.8.1 Field data acquisition protocols

Information acquired from the monitoring relays will include unsolicited data and solicited or requested data:

Unsolicited data is streamed or multicast by the relay without being polled by the central controller, either periodically (with a heartbeat), or when a specified condition or event is observed by the relay. Unsolicited data from the SIPS monitoring and mitigation relays may include any of the following:

- Binary status and alarm points multicast via Ethernet Layer 2 (not TCP/IP; no destination address) in IEC 61850 GOOSE format (1 s heartbeat, which accelerates to a GOOSE packet every 2–4 ms if any point changes, and then gradually slows down again to 1 s).
- Analog values of watts, VARs, current, and voltage multicast in IEC 61850 GOOSE format (fixed heartbeat of 2 to 4 values per second).
- As an alternate form of analog values – synchrophasor measurements multicast in IEC 61850 GOOSE format (same heartbeat as for analog values via GOOSE).

- As an alternate form of analog values – synchrophasor measurements streamed in IEEE C37.118 Synchrophasor Standard format (fixed heartbeat from 2/s to 50/s).

Solicited data is sent from the monitoring or mitigation relays in response to a specific request from the central controller or associated servers on the control center network. It may include any of the following:

- Oscillography records in IEEE C37.111/IEC 60255-25 COMTRADE format, via vendor-specific protocol or FTP over Ethernet TCP/IP (when requested by central server after a system event).
- Event records from SIPS monitoring and mitigating relays, in vendor-specific format handled by vendor-supplied client software, over Ethernet TCP/IP (when requested by central server after event).
- Relay setting or parameter files in vendor-specific format handled by vendor-supplied client software, operating on Ethernet TCP/IP (when requested).

3.8.2 Mixing protocols

A significant feature of Ethernet LANs and WANs is that they can carry mixed traffic types – there is no limitation to one protocol as was the case with serial communications of the past. Thus IEC 61850 GOOSE, IEC 61850 client–server object traffic, IEC 60870-5, or DNP3 SCADA type traffic for monitoring and control, vendor-proprietary relay and intelligent electronic device (IED) configuration communications, COMTRADE oscillographic data files, and event files can all be exchanged over the same network. With proper configuration of priority or QoS, the low-speed monitoring and support functions will have no impact on the performance of the critical high-speed traffic. GOOSE messaging in particular uses VLAN and priority features of Ethernet messages to speed through the system without delay by irrelevant or lower priority traffic.

3.9 SIPS monitoring and testing functions

3.9.1 Testing facilities

The SIPS should include facilities for maintenance testing of any particular analytic or hardware element:

- Isolate trip outputs of mitigating relay(s);
- Isolate and inject analog and status test signals to relay;
- A test switch is to be provided to cause published GOOSE messages and other outputs from a relay to be flagged as *test mode*;
- A separate test switch is to be provided to cause subscribed GOOSE messages and other control inputs to be flagged as *test mode*;
- Test mode status selections shall be alarmed at the central controller (using additional GOOSE status bits).

- At the central controller, individual algorithms or analytics can be placed in test mode while others remain in service.
- If a particular analytic is placed in test mode at the central controller, its mitigating action messages are marked as test mode.

Also, at the field mitigating relay:

- Provide a trip test button or input that causes the relay to send a test status change to the central controller, which is echoed back as a trip output. This verifies communications and the trip output together.

Also, at each central controller:

- Include a means of injecting a test data stream that can stimulate analytics in lieu of a field relay.

These features support annual or periodic testing of all SIPS functions. However, in spite of these features, the SIPS shall be designed for continuous condition monitoring that minimizes the requirement for periodic testing, as explained next.

3.9.2 Monitoring and condition-based maintenance (CBM)

A critical design conception and specification task is for maintainability. Once commissioned, the SIPS is a critical resource that is not readily taken out of service for reconfiguration or testing.

The maintenance strategy is based on condition monitoring rather than testing. In principle, condition monitoring is possible for all components of the proposed SIPS except for the trip output contacts and breakers. This is a key benefit of a control system based on microprocessor relays with data communications paths to the central controller processors, plus constant live data processing, and a heartbeat of periodic unsolicited messages through all communications paths and directions. If condition of system components is fully monitored in an overlapping fashion (no dead unchecked gaps or connections between components), no other testing is required to assure system and equipment health. It is recommended against testing of components and systems that are already known to be working through self-monitoring, since testing introduces risk of human error – a major cause of false trips in the experience of the industry.

The system features to support CBM are:

- For independence from other systems, the SIPS system should be designed to gather all monitoring or self-verification information through its own device processes and communication infrastructure. Any alarms connected to SCADA are optional.
- Central controllers shall have self-monitoring and alarming capability for all components that are critical to SIPS operation, with overlapping of checks and no unverified parts. The design shall document its immunity to nonalarmed malfunctions such as controllers stuck in program loops.

- Catastrophic failures of controllers, including power supply failures, shall be alarmed by other connected devices or systems that lose communications with the failed controller. All monitoring and alarming paths shall themselves be monitored.
- The SIPS relays shall have output contacts for alarming of relay failures through SCADA RTUs, so it can be connected this additional relay monitoring if desired (although failures will be detected by the SIPS controllers).
- Maintenance or failure alarms received or deduced by condition monitoring applications in central controllers, including input selection comparisons of relay measurement values, shall be presented to EMS operators with specifics for action through the SIPS controller displays and through links to SCADA/EMS.
- The presentation processes for malfunction alarms to EMS operators or others who respond to failures shall themselves be monitored for correct and continuous operation.
- The controllers shall generate alarms with specific information for automatic restart or failover operations even if the system or component restores normal operation.
- The controllers should provide the administrator or operator with the capability to restart or reboot central controller processors and other system components that are capable of rebooting. The restarting capability shall also be available from EMS when interfaced to the SIPS.
- Central processors, monitoring relays, and mitigation relays are programmed to monitor integrity of communications messages and to report missing, delayed, or corrupted GOOSE or streamed packet statistics to measure performance of communications channels.
- Health and performance of Ethernet routers or switches shall additionally be monitored via alarms cross connected to relays in addition to any substation alarm annunciator. Ethernet networking devices can also be managed over the network from servers at the central controllers for handling relay data.
- Each SIPS component shall verify an appropriate match of the GOOSE configuration revision (ConfRev) for received messages before acting on them. This avoids misoperations during deployment phases of an updated GOOSE configuration across the SCE system.
- The central controllers shall be compatible with 61850-6 compatible SCL tools for verifying and documenting the connections among specific incoming GOOSE messages from monitoring relays, the SIPS analytics, mitigation messages, all self-monitoring and alarming messages and alarm processing functions.
- Trip circuit monitor (TCM) reports from mitigating relays shall be used to verify the continuity and energization of all mitigating trip circuits and report failures (if compatible with existing trip circuit connections).
- Any change of settings or parameters in any relay or device shall cause an alarm that can be acknowledged by the manager of SIPS operation.

- The settings or parameters of SIPS relays, communications devices, and central controllers shall be periodically (annually) verified against a version controlled archive of settings and after any maintenance activities that can impact settings.

3.9.3 Configuration management

Central controllers and relays all require configuration by large records of parameters, and the parameterization of all the components must match precisely or the system is may misoperate. Notably, GOOSE message point assignments must be properly identified and consistent among the system devices.

Settings should be managed by a version control system that insures the correct parameters are developed and used in maintenance replacement or system updating work. Such a system is characterized by:

- A workflow management process for settings development – creation and entry, validation by peer review, testing review, posting for installation (pending), as-left, obsolete versions.
- A closed-loop process that insures the pending settings are the settings programmed into the correct field device, which has the expected versions of firmware and hardware.
- A facility for periodically comparing settings records uploaded from operating equipment with the archived as-left standard records.
- No settings or parameters left out of the process. Include relays, central controller operating systems and programs, networking device parameters.
- No dependence on the good behavior or information handling perfection of busy maintenance or engineering personnel.

Use a version-controlled management archive with development, signoff, deployment, in-service checking processes, and documentation to maintain all the settings and configuration for:

- Monitoring relays;
- Mitigation relays;
- Central controller configuration and any operating system subject to updating;
- Analytics or algorithm programming;
- Communications configuration (binary, status, control, alarm, analog point assignments and relationships, configuration revision number);
- Communications with and presentation to EMS if used;
- Any Ethernet switches and routers.

The system or a maintenance activity shall directly read (not write) settings, firmware/software versions, and related hardware data from SIPS components over communications channels and verify against the managed archive.

Managed settings verification is required to support the CBM program in which end-to-end periodic tests of each individual SIPS analytic are not routinely performed. In a CBM program, most of the system is *not* functionally tested periodically after a commissioning test.

3.10 Example of SIPS application based on PMU technology

The objective of this SIPS example is to present automated monitoring and control (AMC) based on PMU technology of transmission grid with integrated not only distributed generation (wind power plants (WPPs)) but also hydro power plants (necessary for regulation) and basic thermo power plants (coil) [17]. AMC enables control of a part of transmission system with an aim to mitigate possible disturbances and, if necessary, to establish islanding operating condition to keep consumers reliably supplied by electrical energy. Automated operation of a part of transmission system is based on synchronized measurements and mitigating devices (IED) deployed in the transmission system based on a priori off-line case studies. Developed architecture of AMC is founded on anticipated functional requirements developed through system studies and through industry experience and also similar solutions deployed in transmission power system through SIPS. Especially, requirements for distributed generation will be specified to enable AMC to manage possible congestions and in extreme case form an island operation. The multifunctional system acquires a flexible and general architecture based on centralized control and data communications. The flexibility of AMC assumes programming with a range of specific algorithms that might be created as study and development continues.

Most operators run their systems on the principle of N-1 criteria. However, faced with new market-oriented conditions in the electrical power system (EPS) operation, where a minimal possibility for influencing the level of electricity production, especially from renewable sources, leads systems towards situations which were not originally intended and designed for, force operators to operate, use and analyze the systems based on N-0 criteria accepting the higher risk [18].

3.10.1 Operational and influence analysis of WPP on the EPS

New operating conditions, which are necessary to be analyzed in order to get good knowledge about functioning of the part of transmission network and achieving the possible savings in electricity transmission, have been defined by the construction of WPP and specific connections of other electrical energy sources within the specified area.

A mathematical model of the part of transmission network was designed for the analysis of WPP Vratarusa regarding the settings of protection functions in the area. The model is based and verified on the basis of synchronized phasor measurements of voltage and current in the system. The mathematical model was used for the analysis of normal, extraordinary, and breakdown operating conditions and their impact on the operation of relay protection and stability of the whole system. The analysis included simulations of dynamic changes in the critical points of the power system, such as line outages, power outages, load changes, power swings, transient and permanent short circuits.

3.10.2 Problems regarding WPP Vratarusa

The structure of the electrical energy produced in Croatia during the year is constantly changing and adapting. Thus, in one part of the year dominates thermal while in the second, a relatively shorter period dominates hydro generation. Import of energy varies depending on total production and the needs of the power system. All major thermal power plants are located in the continental part of the country and largest hydroelectric sources are located in the south of the country. At the time of good hydrological conditions, energy is mainly transporting from south to north of Croatia. A WPP Vratarusa was built in this corridor that under suitable wind conditions can cause some instability in the system (e.g., overload of transmission lines).

WPP Vratarusa has 14 generators with a total output of 42 MW and is connected to the 110 kV network between the two hydroelectric power plants: HPP Senj (210 MW) and HPP Vinodol (84 MW). Energy generated in HPP Senj and WPP Vratarusa is mostly transmitted via 110 kV transmission line Vratarusa-Crikvenica. This, already highly loaded transmission line, under certain operating conditions, may be switched off from the system due to overloading or actions provided by relay protection devices. Transmission congestion problem can be temporarily solved by limiting the production of HPP Senj, but then there is a risk of overflowing of already full accumulations of this power plant. This fact and connection of WPP in that area are the basic problems of placement their energy into the grid.

3.10.3 N-1 Analysis regarding the specific disturbances in EPS

The analysis of 220 kV transmission line outage Senj–Melina and 220/110 kV transformer outage in HE Senj was carried out using the N-1 criteria as a serious disorders in the system around WPP Vratarusa. All analysis was carried out for the case of maximum generation of hydro power plants Senj and Vinodol when the WPP Vratarusa was in and out of the operation. Analysis of the power flow with N-1 criteria shows that:

- in case of maximum engagement of HPP Senj (210 MW) and WPP Vratarusa (42 MW), N-1 criterion is not met due to failure of 220 kV transmission line Senj–Melina;
- in case of maximum engagement of HPP Senj (210 MW) and WPP Vratarusa (42 MW), N-1 criterion is also not met due to failure of 220/110 kV transformer in HPP Senj;
- reducing the generation of HPP Senj (in the case when the WPP Vratarusa is on the network) can reduce N-1 criterion within the allowable limits;
- N-1 criterion is met when WPP Vratarusa is out of operation for selected scenarios;
- it is possible to satisfy the N-1 criterion by separation of 110 kV bus bars in HPPs Senj and Vinodol.

Figure 3.9 shows the schema of the observed part of the smart transmission system.

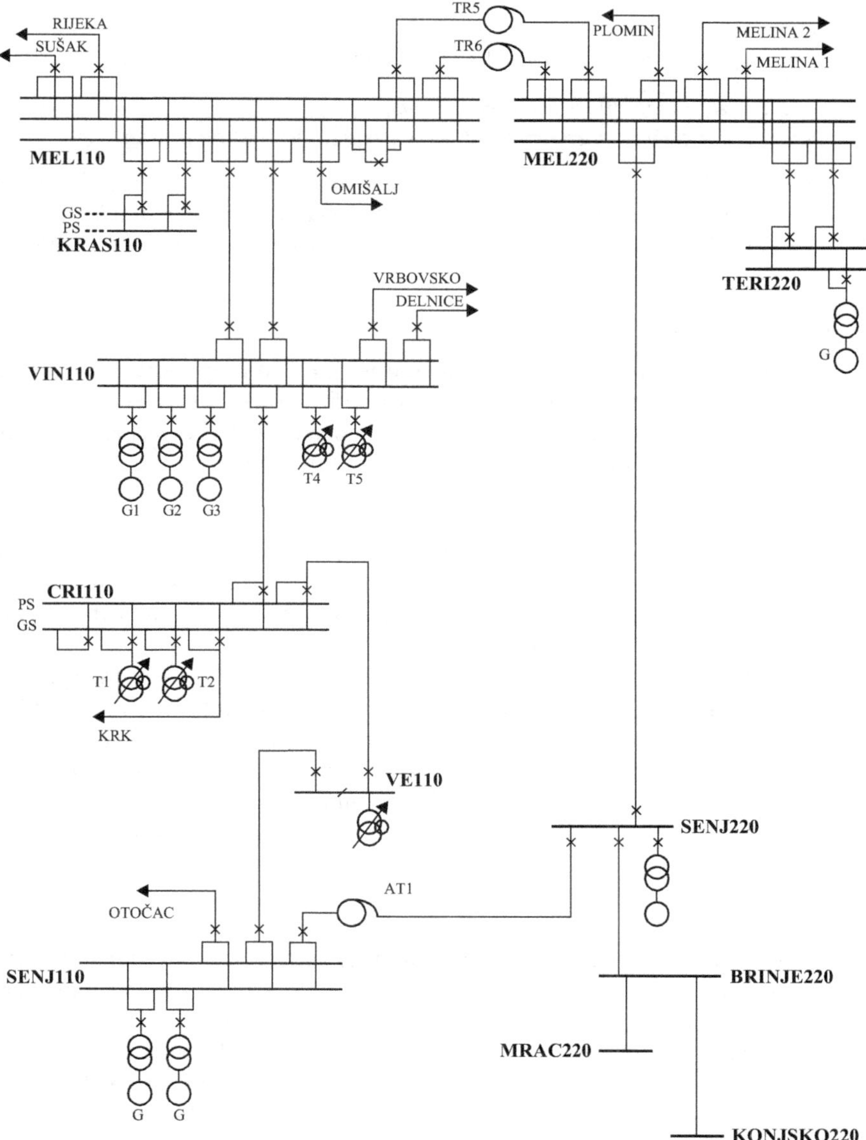

Figure 3.9 Schema of the observed smart transmission system

3.10.4 Conceptual design of automatic control and monitoring

This example describes algorithms and techniques that have been used for congestion management and possible islanding operation.

AMC operation

Therefore the proposed AMC shall be able to accomplish the following:

- *Detect and manage possible congestions;*
- *Detect the external emergency for which islanding will be formed*

  Detection shall include following measurements: load flow patterns, frequency or change of frequency at specific locations, bus voltage magnitude relationships and bus voltage angular relationships across the observed distribution network.

- *Load shedding*

  One real benefit of establishing a central island is to keep critical loads energized. It is better to drop only the lower priority loads. This selective tripping can be carried out only by the distribution operator based on a load shedding signal from the AMC mitigation relay at the substation.

- *Detection and islanding*

  Some indicators are abnormal load flows out of or across the observed distribution network, depressed voltage, protective relay disconnections of ties to the external system, and frequency decline which are programmed in multi-factor authentication of an emergency.

  Based on indicators, best performance for the AMC shall provide synchrophasor measurements. It can show gradual divergence of the observed network from external system locations and from normal operating states. Relays readily available today can deliver synchrophasor values via Ethernet communications, either in the format of communications part of IEEE Synchrophasor Standard C37.118, or as time-stamped analog values in IEC 61850 GOOSE messages.

- *Restoration*

  Because the major outage that triggered the islanding might have many causes and system wide impacts, automated restoration will not be recommended. Once the island is formed, it seems likely that operators will want to further tune operation in the island, and later to restore normal configuration manually. The role of the AMC is then to monitor the observed network during the restoration process and to shed loads to avoid possible problems.

AMC Solution

Most of the SIPS systems that are working around the world are designed to reduce special congestions. These AMC send measured values and switching statuses change from different locations to the controller that starts mitigation activity if the measured values and switching statuses coincide with predefined conditions that require mitigation activities. If the controller determines that there is such a condition, it sends the appropriate control command towards substations where the

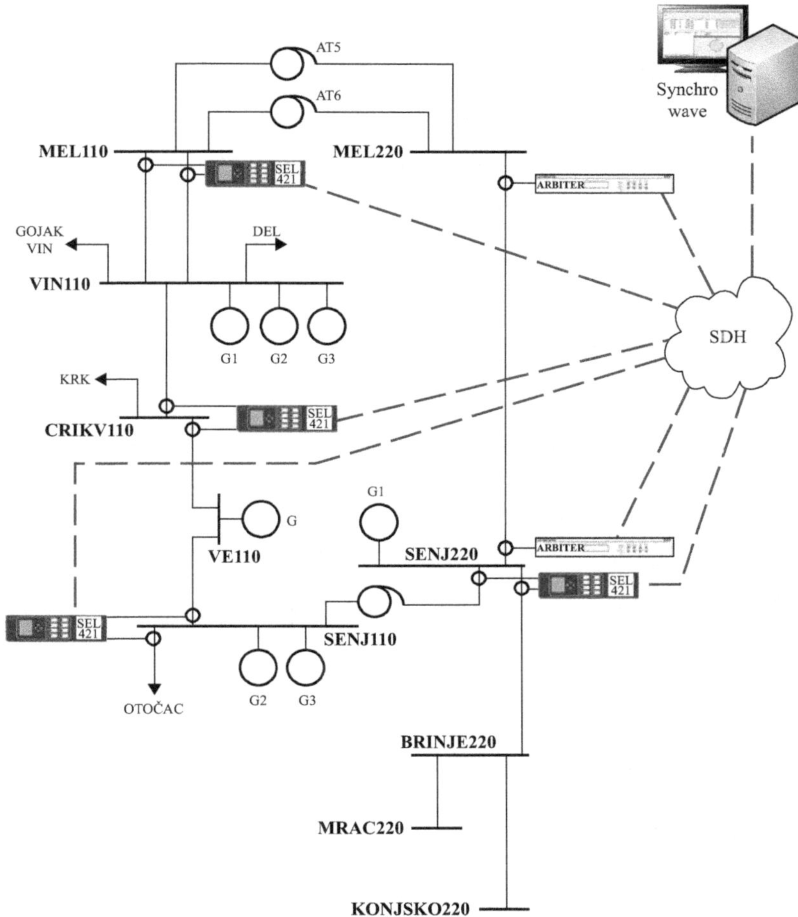

Figure 3.10 AMC system overview

required mitigation activates. Figure 3.10 shows positions where IED are deployed, based on a priori case studies.

When there is need to include a various range of AMC strategies in one system, it is more efficient to use a centralized architecture. In the centralized approach, the platform for decision-making is a combination of relays, computers (servers), and PLCs installed in the control center in which all communication paths converge. Relays in substations will serve as a source of analog measurements and as an indication of the binary statuses of electrical equipment. These same relays also represent the final output to the control of electrical equipment. The difference with standard control functions is that the relays do not make local decisions about the state of the power system in substations. Their task is to pass the

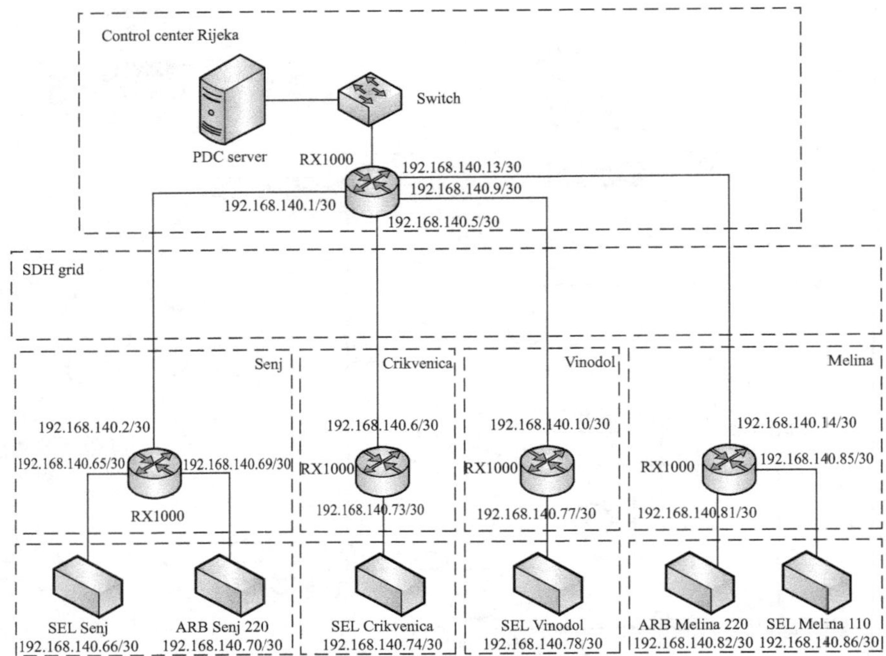

Figure 3.11 Proposed AMC architecture

measurements and switching statuses to central controller so that the central controller can consider data from the entire network and make decisions on activities for entire power system.

3.10.5 System architecture overview

The AMC system architecture, which is schematically depicted in Figure 3.11, is based on synchronized measurement (PMU) and mitigating relays. Measurements are concentrated in central controller and processed. Resulting signals are send to mitigating relays in order to disconnect the observed transmission network from the external system and carry out congestion management procedures or to set an island mode of operation. The communications channels for critical binary values are high-speed digital communication channels similar to those used for pilot or unit protection of transmission lines.

For AMC that has a unique purpose the most efficient way is to place platform that makes decision about analog measurements warning levels in the substation where the analog measurements are going to be made, since the transfer of these measurements would require complex high-speed communication channels. For example, the line terminal relay of the substation reveals when the line load exceeds the nominal value calculated in a priori study. An indication that the load exceeded the target is transferred to a distant point as a simple binary signal where the controller must detect congestion, such as line or generator outage. In this kind of AMC, there is no need for transferring any high-speed analog data and AMC can react to destination within less than 100 ms.

However, as soon as AMC prompts for comparisons of analog values with multiple distant points, or if it requires flexibility and strategy evaluation for analog values, then these values must be transferred between distant points that greatly extends the reaction time of AMC.

Even when a project focuses on the maximum transfer speed of all critical information, the binary signals can always transfer faster and can update more frequently than the analog values. Therefore, the AMC project logic and analytics are frequently using analog values only for data processing and not all the while for direct activities. In some conventional AMC projects, monitoring of analog measurements does not require the rapid detection of sudden changes in values. Communication channels for critical binary values are then fast digital communication channels similar to those used for local protection of transmission lines (remote line trip).

This kind of strategy may not always be adequate. If the congestion consists of element overloading without the specific cause than the analog values that surround the element must be processed in real time to make a decision on the need for activity. The subject AMC includes overloading measurements and proposed project plans to transfer analog values by a relatively high rate of speed and updates.

AMC has the ability to fully implement the independent functions in relation to the EMS or SCADA and existing protective relays and control systems in the substations. This particularly applies to the following:

- individual view for the operator associated to each AMC controller and the ability to display the operating status of monitored transmission network including the status of switches and power flows in parallel with the EMS (graphical representation is not specified although it is possible);
- Some switches can be managed with AMC, mostly the executive ones.
- The speed of updating and refreshing the data is far better than those in the EMS.

AMC which is related to the EMS would be far slower and more vulnerable to a single-point communication or the failures of standard remote terminal units, though its implementation would be much cheaper. Nonredundant load measurements are available in ranges from a few seconds to few minutes. Also, AMC becomes associated with the possible EMS replacing projects and the EMS maintenance and updating activities.

Associating the existing system relays to the AMC questions the possibility of transferring status signals and analog values with AMC, and the question of used protocols. Maintenance, testing, and updating of any protection system or AMC function becomes a risk as both the one and the other affects any activity.

Although AMC can work independently, it is specified that it provides possibility to communicate with operators in the control center while working in their normal working environment.

References

[1] UCTE. *Final Report of the Investigation Committee on the 28 September 2003 Blackout in Italy.* UCTE Report – April 2004.

[2] Andersson G., Donalek P., Farmer R., *et al.* "Causes of the 2003 major system blackouts in North America and Europe, and recommended means to improve system dynamic performance." *IEEE Transactions on Power Systems*, vol. 20, no. 4, November 2005.

[3] Terzija V., Valverde G., Cai D., *et al.* "Wide area monitoring, protection and control of future electric power networks." *Proceedings of the IEEE, Special Issue – Network Systems Engineering: Meeting The Energy & Environmental Dream*; January 2011, Volume 99, Number 1, ISSN 0018-9219, coden IEEPAD.

[4] Begović M., Madani V., Novosel D. "System integrity protection schemes (SIPS)." *Bulk Power System Dynamics and Control – VII. Revitalizing Operational Reliability, 2007 iREP Symposium*, Charleston, SC, USA, pp. 1–6, 19–24 August 2007.

[5] Novosel D., Begović M., Madani V. "Shedding light on blackouts." *IEEE Power and Energy Magazine*, Volume 2, Issue 1, pp. 32–43.

[6] Madani V., Novosel D., Horowitz S., *et al.* "IEEE PSRC Report on Global Industry Experiences With System Integrity Protection Schemes (SIPS)." *IEEE Transactions on Power Delivery*, Volume 25, Issue 4, pp. 2143–2155, 11 October 2009.

[7] Madani V., Taylor E., Erwin D., Meklin A., Adamiak M. "High-speed control scheme to prevent instability of a large multi-unit power plant." *60th Annual Conference for Protective Relay Engineers*, College Station, TX, pp. 271–282, 27–29 March 2007.

[8] Begović M., Madani V., Novosel D. "System integrity protection schemes (SIPS)." *2007 iREP Symposium Bulk Power System Dynamics and Control – VII. Revitalizing Operational Reliability*, SC: Charleston, IEEE, pp. 1–6, 19–24 August 2007.

[9] Plavšić T., Skok S., Ivanković I. "Special protection scheme for operation of Central Zagreb transmission system." *Engineering Review*. vol 31, No. 1 ISSN: 1330–9587, July 2011.

[10] Phadke A.G., Thorp J.S. *Synchronized Phasor Measurements and Their Applications*. New York: Springer, 2008.

[11] Skok, S., Šturlić, I., Matica, R. "Multipurpose open system architecture model of wide area monitoring." *PowerTech, 2009 IEEE Bucharest*, June 28 2009–July 2 2009.

[12] Bertsch J., Zima M., Suranyi A., Carnal C., Rehtanz C. "Experiences with and perspective of the system for wide area monitoring of power systems." *CIGRE/PES Quality and Security of of Electric Power Delivery System*. 2003; *CIGRE/IEEE PES International Symposium*, IEEE, pp. 5–9, 8–10 October 2003, pp. 5–9.

[13] Zhou M.,Centeno V.A, Thorp J.S., Phadke A.G. "An alternative for including phasor measurements in state estimators." *IEEE Transactions on Power Systems*, vol. 21, no. 4, November 2006, pp. 1930–1937.

[14] Skok S., Matica R., Šturlić I. "Enhanced open architecture of phasor data concentrator", *European Transactions on Electrical Power*, 2010, Published online in Wiley Online Library (wileyonlinelibrary.com). DOI: 10.1002/etep.527

[15] Patel M.Y., Girgis A.A. "Two-level state estimation for multi-area power system," *Power Engineering Society General Meeting*, 2007. FL: Tampa, *IEEE*, 24–28 June 2007, pp. 1–6.

[16] Skok S., Ivanković I., Cerina Z. "Applications based on PMU technology for improved power system utilization." *Power Engineering Society General Meeting*, 2007. FL: Tampa, *IEEE*, 24–28 June 2007.

[17] Skok S., Ivanković I., Frlan K., Zbunjak Z. "Monitoring and control of smart transmission grid based on synchronized measurements." *APAP 2011*, 16–20 October, Beijing, China.

[18] Novosel D., Vu K., Centeno V., Skok S., Begović M. "Benefits of synchronized measurement technology for power-grid applications." *HICSS 2007*, 3–6 January 2007, Hawaii.

Chapter 4

New methodologies for large-scale power system dynamic analysis

Aleksandar M. Stanković[1] *and Andrija T. Sarić*[2]

4.1 Introduction

In contemporary power system practice, the stability limits are typically computed off-line and stored in databases to be monitored by system operators (dispatchers) in the real-time environment. Several sources of uncertainty affect such computations and consequently reasonable stability margins must be taken into account when determining operation limits. Despite these precautions, unplanned outages and planned switching actions (such as topology control [TC] actions) may cause operational conditions not considered at operation planning stages, and consequently system operators are left with no pertinent stability information. Online stability assessment has been proposed as an additional line of defense in which stability limits are computed based on the actual power system condition, which decreases the uncertainty, thus providing more accurate stability operation limits. The online stability assessment can be performed for the (real-time) operating point only, or additionally for a region around this condition [1].

Power system stability analysis requires the evaluation of several different aspects such as thermal limits, steady-state and transient voltage levels, small-signal stability, transient stability [2–8]. However, due to the computational complexity, these simulations for large-scale power systems remain a challenging task. Two main research strategies have been proposed and adopted in the industry to deal with the computational complexity [1].

One is to simplify the problem by using *reduced network and/or dynamic models* and to compute stability indices using faster calculations. Such indices are not expected to be exceedingly accurate, but to assess the proximity to steady-state, dynamic, or transient instability. Different approaches have been developed along these lines, such as transient energy functions [9, 10], single machine equivalent [11], steady-state stability indices [12].

[1]Department of Electrical Engineering and Computer Science, Tufts University, Medford, MA, USA
[2]Department for Power, Electronic and Communication Engineering, Faculty of Technical Sciences, University of Novi Sad, Novi Sad, Serbia

The other research strategy is to use *full time-domain simulations* with detailed models, tailored simulation algorithms, and parallel processing. Detailed models are used (at least) for the main area of interest, and dynamic equivalents can be used for the remainder [13, 14]. Stability programs designed for large-disturbance (transient) stability studies simulate the system response in time domain. The simulations are normally limited to a short duration (usually a few seconds). Time-domain simulations can tell whether the system is transiently stable or not, largely in a binary fashion [1]. However, the main disadvantage of full time-domain simulations, in particular when detailed models of dynamic elements are used, is the inherently high computational cost [15]. Instability in the initial period following a large disturbance is generally due to insufficient synchronizing torque between the interconnected generators. Automatic control components (e.g., fast-acting excitation control, FACTS-based control) help to improve the first swing stability by increasing the synchronizing torque. However, in the process they often reduce the damping torque, sometimes even rendering the overall damping negative, thereby aiding oscillatory instability. With the growth of interconnections and application of advanced control equipment, consideration of proper damping of oscillations became more important. In a power system capable of withstanding the initial shock of the large disturbance, as evidenced by first swing stability, oscillations could continue at reduced amplitude for a while, only to increase later and eventually cause cascading line tripping and possibly system separation (islanding). This type of instability can manifest itself not only following a major disturbance but also following a sudden small change in system condition (e.g., a moderate amount of load tripping, a sudden addition of a large load, tripping of a minor transmission line) [5].

Conventional stability programs designed for large-disturbance stability studies can, in principle, be used to study the above phenomenon by simulating longer time intervals. This approach can, however, be impractical and the result may not be conclusive.

Immediately following a small disturbance, or following a large disturbance after the system has survived the initial shock (i.e., the system is first-swing stable) and entered a state of oscillation, the system nonlinearities do not play a decisive role. The power system can therefore be linearized about the equilibrium point and useful information on the system small-disturbance performance can be obtained from the linearized model.

Although there are several methods of obtaining stability information from a linearized system, a state-space approach is preferable since it provides additional discipline in modeling and can handle complex power systems [2–9, 11, 16].

Our experience with eigenvalue sensitivity studies [17] is that they can be effective in identifying and quantifying the power system behavior in both dynamic and steady-state conditions. Two important developments serve as cornerstones for our large-scale studies: (1) sparsity-preserving methodology [18] and (2) the ability to identify eigenvalues most affected by a given power system change [19]. In industrial practice, eigenvalues and other small-signal characterizations need to be

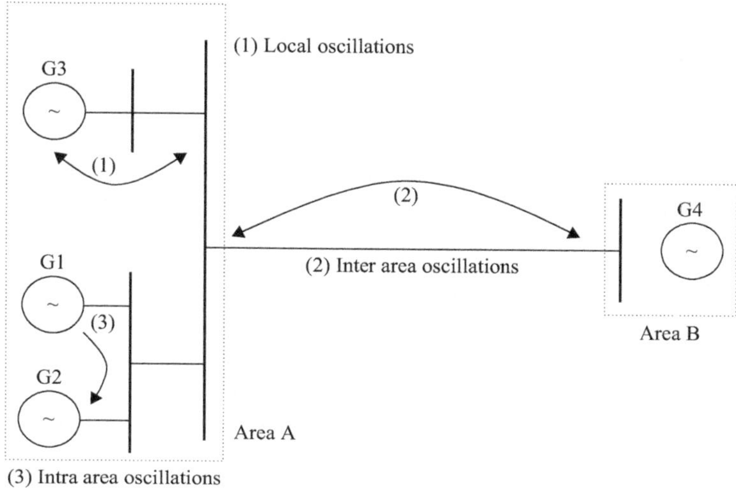

Figure 4.1 Types (modes) of oscillations in power systems

complemented with appropriate quantification of large-signal properties such as the transient stability simulations.

The small-signal stability study corresponds to the stability considerations for chosen "nominal" operating conditions (around a selected steady-state condition) of the power system in the presence of normal low-amplitude fluctuations of electrical or mechanical variables (e.g., mechanical torque, governor power reference input, or automatic voltage regulator [AVR] input reference).

Low-frequency electromechanical oscillations occur, for example, when existing generation/load areas are connected to other similar areas by relatively weak transmission lines (called tie-lines).

In many power system cases, instability and eventual loss of synchronism are initiated by some disturbance in the system, leading to an oscillatory behavior. If not damped, the oscillations may eventually build up and lead to component tripping and even the system blackout. Oscillations are undesirable because they limit power transfers on transmission lines and potentially induce stress in the mechanical portions of generators. Root causes of local and inter area oscillations are difficult to pinpoint.

The electromechanical oscillations can be classified into two broad groups, illustrated in Figure 4.1 [4, 6–8, 16, 20]:

- *Local- (plant-) mode oscillations* are the most commonly encountered and are associated with oscillations between a remotely located power station and the remainder of the power system. These oscillations are usually caused by the action of the AVRs of generating units, operating at high output and feeding into weak-transmission networks. The local mode oscillations typically have natural frequencies in the range of 1–2 Hz. They can be readily damped by using supplementary control of excitation systems in the form of power system stabilizers (PSSs).

- *Inter area mode oscillations* are associated with machines in one part of the power system, oscillating against machines in other parts of the power system. They are caused by two or more *groups* of closely coupled machines that are interconnected by weak tie-lines. This mode of oscillation can appear following a major disturbance in any of the areas separated by weak tie-lines. The natural frequency of these oscillations is typically in the frequency range below 1 Hz. This is lower than local- (plant-) mode oscillations because of higher combined inertias and relatively weak tie-lines. The characteristics of these oscillations are complex and in some respects significantly differ from the characteristics of other oscillation modes [6, 16].
- *Intra area mode oscillations* can be divided into two characteristic subtypes: *torsional mode oscillations* and *control mode oscillations*. *Torsional mode oscillations* are associated with the turbine-generator rotational (mechanical) components. There have been several instances of torsional mode instability due to interactions with controls, including generating unit excitation and prime mover controls [6]. *Control mode oscillations* are associated with the controls of generating units and other dynamic equipment (excitation systems, prime movers, static Var compensators, high-voltage DC (HVDC) converters, etc.). Poorly tuned controls are the usual causes of instability of control modes. The stabilizers used the shaft-speed signals with torsional filters, or an alternative form of stabilizer without torsional filters [6].

Although it is possible to identify oscillation modes from time-domain simulations [21 and therein references], it can often be done more accurately from analysis of the linearized system, which also reveals coupling between modes.

Voltage stability study investigates the voltage problems from different aspects of power system operation conditions and recommends the appropriate solutions as protection and preventive control actions. IEEE–CIGRE Joint Task Force on Stability Terms and Definitions refers the voltage stability as "the ability of a power system to maintain steady voltages at all buses in the system after being subjected to a disturbance from a given initial operating condition" [22]. In general, the inability of the power system to supply the required demand (especially reactive) leads to voltage instability (voltage collapse). *Voltage collapse is the process by which voltage instability leads to a loss of voltage in a significant part of the power system.*

Voltage stability studies have two principal aspects: *slow/small disturbance* (long term, with voltage collapse in minutes to hours) and *fast/large disturbance* (short term, with voltage collapse in the order of fractions of a second to a few seconds). They are concerned with the ability of a power system to maintain acceptable voltages at all buses, according to $(n-1)$ and/or $(n-2)$ concepts of security.

Short-term voltage stability problems are usually associated with the rapid response of voltage controllers (e.g., generators' AVR) and power electronic converters, such as those encountered in flexible AC transmission system or FACTS controllers and HVDC links. In the case of voltage regulators, the voltage instability is usually related to inappropriate tuning of the power system controllers. Voltage stability in converters, on the other hand, is associated with commutation issues in the electronic switches, particularly when these converters are connected to "weak" AC power systems, i.e., power systems with poor reactive power support.

Dynamic voltage stability (DVS) has been gaining increasing importance because today's power systems are operated closer to their VS limits [2, 3, 23]. Methods used to analyze the DVS problem have theoretical foundations in the structural stability theory of dynamical systems [24, 25]. Nonlinear phenomena have been shown to be directly associated with voltage stability problems and differential–algebraic equations (DAEs)-based model of the power system. This model is often linearized at an equilibrium point, so that scalable and versatile methods of linear system theory can be deployed.

In power systems, several bifurcation types are typically of concern [2–9, 11, 26]:

- *Saddle-node (fold) bifurcations* (SNBs) are related with loading margins. Singularity of the DAEs augmented matrix associated with the linearized DAEs model is an indication of the SNB. This point can be identified using either direct or continuation methods [27–31]. Direct methods find SNB by solving an augmented system of equations.
- *Hopf bifurcations* (HBs) are related with oscillatory stability. The HB occurs when pair of complex conjugate eigenvalues move from negative to the positive real parts by crossing the imaginary axis under parametric variations [32–37].
- *Singularity-induced bifurcations* (SIBs) are related with singularities of power flow-based algebraic equations (AEs). Several authors [32, 38, 39] have shown that the SIBs occurs when system equilibria encounter the singularity manifold and it refers to a stability change owing to one of the eigenvalues of the reduced power flow-based Jacobian matrix associated with equilibrium diverging to infinity (algebraic singularities).
- *Limit-induced bifurcations* (LIBs). A type of instantaneous changes in the eigenvalues of reduced system matrix similar to SIBs is observed in the case of LIBs that occur when limits of the control range or other differential/algebraic variables are reached [39].

For load/generation increments (used for DVS tracing) and switching list for TC actions, the power system must stay in the feasible region in transient and long-term time scales. Quasi steady-state model is incapable of capturing the instabilities where the long-term unstable evolution triggers a short-term instability [40–42].

The first limitation lies in the implicit assumption that the neglected short-term dynamics are stable. After a significant change (such as a TC action), the power system may lose stability in the short-term frame (e.g., within 10 s after the change) and hence not enter in the long-term phase simulated under the quasi steady-state approximation [28].

The second limitation is linked to the discrete events, typical for TC actions. These events may trigger controls with great impact on the power system long-term evolution. This means that the system dynamics depends on a sequence of controls, and hence may not be correctly identified from the simplified quasi steady-state model [43].

In this chapter we focus on the scalability issues of calculations needed to ascertain various types of bifurcations. In the case of large-scale power systems (thousands of buses, tens of thousands of eigenvalues), it is simply not feasible to perform repeated calculations of all eigenvalues, or to perform full optimization-based algorithms with computer resources available today [15, 44, 45]. We are thus interested in algorithms that remain tractable for such large-scale power systems by focusing on relevant portions of the eigenvalue spectra.

The methodologies described in this chapter are verified on two test examples: small-scale (New England, 39-bus, 46-branch and 10-generator) and large-scale and real-world (Pennsylvania-Jersey-Maryland (PJM), 13709-bus, 18285-branch and 2532-generator).

4.2 Dynamic model

The typical analytical approach for small-signal studies is a multimachine linearized analysis that computes the eigenvalues of the system matrix and finds machines that contribute to a particular eigenvalue, causing the instability (both local and inter area oscillations can be studied in such framework). Formally, instability occurs when a pair of complex-conjugate eigenvalues crosses to the right-half plane; in practical terms, a system is deemed unstable when the damping ratio (that will be defined in Section 4.3) of oscillations is insufficient. Small-signal studies are connected to transient stability analysis through the need to determine the system equilibrium condition. During this model initialization values of variables are calculated for interfaces of each component model.

To develop a power system dynamic simulation model appropriate for small-signal and dynamic stability studies, the equations used for modeling various dynamic components (such as generators, exciters, governors) are written as [46]:

$$\dot{\boldsymbol{x}}_d = \boldsymbol{f}(\boldsymbol{x}_d, \underline{\boldsymbol{V}}_d) \tag{4.1}$$

$$\underline{\boldsymbol{I}}_d = \boldsymbol{g}'(\boldsymbol{x}_d, \underline{\boldsymbol{V}}_d) \tag{4.2}$$

Assuming that there are $d = 1, 2, \cdots, N$ one-port dynamic devices in the power system and they are connected to buses 1 to N, we obtain following a set of nonlinear DAEs for time-domain simulations:

$$\dot{\boldsymbol{x}} = \boldsymbol{f}(\boldsymbol{x}, \underline{\boldsymbol{V}}) \tag{4.3}$$

$$\boldsymbol{0} = \boldsymbol{g}'(\boldsymbol{x}, \underline{\boldsymbol{V}}) - \underline{\boldsymbol{Y}}'_{\mathrm{BUS}}\underline{\boldsymbol{V}} = \boldsymbol{g}(\boldsymbol{x}, \underline{\boldsymbol{V}}) \tag{4.4}$$

where

$\boldsymbol{x} \in \Re^{n_x}$ – n_x-dimensional vector of state variables, which represents the dynamics of generators, turbines, exciters, loads, and other system components;
$\boldsymbol{f}$ – n_x-dimensional set of differential equations;
$\boldsymbol{g}'$ – $n_{\underline{V}}$-dimensional set of AEs;
$\underline{\boldsymbol{V}} = \begin{bmatrix}\boldsymbol{\theta}^T & \boldsymbol{V}^T\end{bmatrix}^T \in \Re^{n_{\underline{V}}}$ – $n_{\underline{V}}$-dimensional vector of algebraic (complex bus voltage) variables [composed from voltage angle ($\boldsymbol{\theta}$) and voltage magnitude ($\boldsymbol{V}$) component subvectors];
$\underline{\boldsymbol{Y}}'_{\text{BUS}} = \boldsymbol{G}'_{\text{BUS}} + j\boldsymbol{B}'_{\text{BUS}}$ – ($N_{\text{BUS}} \times N_{\text{BUS}}$)-dimensional short-term (transient) analysis-based bus admittance matrix (composed from conductance and susceptance component submatrices, respectively); Note that $\underline{\boldsymbol{Y}}'_{\text{BUS}}$ is different from power flow bus admittance matrix ($\underline{\boldsymbol{Y}}_{\text{BUS}}$) used for calculation of power flow-based Jacobian matrix [$\boldsymbol{J}_{\theta,V}$ in (4.30)], because it contains a network-dependent part and a part corresponding to loads [46]; Also, note that $\underline{\boldsymbol{Y}}_{\text{BUS}}$ is a fixed matrix for load modeled as constant power (with respect to variations in operating conditions);
$\underline{\boldsymbol{I}} = \underline{\boldsymbol{Y}}'_{\text{BUS}}\underline{\boldsymbol{V}}$ – $n_{\underline{V}}$-dimensional set of complex network bus injection equations;
N_{BUS} – total number of system buses;
n_x – total number of the dynamic states (state variables);
$n_{\underline{V}} = 2N_{\text{BUS}}$ – total number of algebraic variables.

After linearization around chosen operating point for differential/algebraic variables, DAE model (4.3), (4.4) can be transformed into incremental form as:

$$\begin{bmatrix}\Delta\dot{\boldsymbol{x}} \\ \boldsymbol{0}\end{bmatrix} = \begin{bmatrix}\boldsymbol{F}_x & \boldsymbol{F}_{\underline{V}} \\ \boldsymbol{G}_x & \boldsymbol{G}_{\underline{V}}\end{bmatrix}\begin{bmatrix}\Delta\boldsymbol{x} \\ \Delta\underline{\boldsymbol{V}}\end{bmatrix}; \quad \begin{bmatrix}\boldsymbol{x}(t_0) = \boldsymbol{x}_0 \\ \underline{\boldsymbol{V}}(t_0) = \underline{\boldsymbol{V}}_0\end{bmatrix} \tag{4.5}$$

or in descriptor form as:

$$\begin{bmatrix}\boldsymbol{1} & \boldsymbol{0} \\ \boldsymbol{0} & \boldsymbol{0}\end{bmatrix}\Delta\dot{\boldsymbol{X}} = \begin{bmatrix}\boldsymbol{F}_X \\ \boldsymbol{G}_X\end{bmatrix}\Delta\boldsymbol{X}; \quad \boldsymbol{X}(t_0) = \boldsymbol{X}_0 \tag{4.6}$$

where:

$\boldsymbol{X} = \begin{bmatrix}\boldsymbol{x} \\ \underline{\boldsymbol{V}}\end{bmatrix} \in \Re^{nX} = \Re^{n_x} \times \Re^{n_{\underline{V}}}$ – ($n_x + n_{\underline{V}}$)-dimensional vector of state and algebraic variables;

$\Delta\boldsymbol{X} = \begin{bmatrix}\Delta\boldsymbol{x} \\ \Delta\underline{\boldsymbol{V}}\end{bmatrix}$;

$\boldsymbol{F}_x = \left.\dfrac{\partial \boldsymbol{f}(\boldsymbol{x}, \underline{\boldsymbol{V}})}{\partial \boldsymbol{x}}\right|_{\boldsymbol{x}_0, \underline{\boldsymbol{V}}_0}$; $\quad \boldsymbol{F}_{\underline{V}} = \left.\frac{\partial \boldsymbol{f}(\boldsymbol{x}, \underline{\boldsymbol{V}})}{\partial \underline{\boldsymbol{V}}}\right|_{\boldsymbol{x}_0, \underline{\boldsymbol{V}}_0}$;

$\boldsymbol{G}_x = \left.\dfrac{\partial \boldsymbol{g}'(\boldsymbol{x}, \underline{\boldsymbol{V}})}{\partial \boldsymbol{x}}\right|_{\boldsymbol{x}_0, \underline{\boldsymbol{V}}_0}$; $\quad \boldsymbol{G}_{\underline{V}} = \left.\frac{\partial \boldsymbol{g}'(\boldsymbol{x}, \underline{\boldsymbol{V}})}{\partial \underline{\boldsymbol{V}}}\right|_{\boldsymbol{x}_0, \underline{\boldsymbol{V}}_0} - \underline{\boldsymbol{Y}}'_{\text{BUS}} = \boldsymbol{G}'_{\underline{V}} - \underline{\boldsymbol{Y}}'_{\text{BUS}}$;

$F_X = \left[F_x \quad F_{\underline{V}} \right]; \quad G_X = \left[G_x \quad G_{\underline{V}} \right];$
$\mathbf{1}$ – n_x-dimensional unity diagonal matrix;
$\mathbf{0}$ – zero matrices (with matching dimensions).

The linearized DAEs model (4.5) can be rewritten in *state space form* emphasizing differential aspects of dynamics:

$$\Delta \dot{x} = A_{\text{sys}} \Delta x \tag{4.7}$$

where:

$$A_{\text{sys}} = F_x - F_{\underline{V}} G_{\underline{V}}^{-1} G_x \tag{4.8}$$

is the state space matrix.

The eigenvalues of F_x are typically all negative, as each device and associated controls are basically decoupled from the others. The matrix $F_{\underline{V}} G_{\underline{V}}^{-1} G_x$ couples each device through the network and algebraic variables (in vector $\underline{V}$), which can be considered *aggregation variables*. Since A_{sys} may show poorly damped or even positive real part eigenvalues, an effective technique for improving or obtaining stability is to reduce, by means of adequate controllers, the effect of $F_{\underline{V}} G_{\underline{V}}^{-1} G_x$ (sometimes labeled the *degradation matrix*) and, in turn, to decouple the destabilizing dynamic connections.

A formal solution of the above state equations can be obtained using the Laplace transforms as:

$$\Delta x(s) = (sI - A_{\text{sys}})^{-1} \Delta x(0) = \frac{\text{Adj}(sI - A_{\text{sys}})}{\text{Det}(sI - A_{\text{sys}})} \Delta x(0) \tag{4.9}$$

where:

I – identity matrix of the same dimensions as matrix A_{sys};
$\text{Det}(sI - A_{\text{sys}})$ – determinant of matrix $(sI - A_{\text{sys}})$;
$\text{Adj}(sI - A_{\text{sys}})$ – adjoint of matrix $(sI - A_{\text{sys}})$.

The poles (eigenvalues of matrix A_{sys}) are roots of the characteristic equation:

$$\text{Det}(sI - A_{\text{sys}}) = 0, \text{ or equivalently } \text{Det}(A_{\text{sys}} - sI) = 0, \tag{4.10}$$

Due to CPU and memory requirements for inversion $G_{\underline{V}}^{-1}$ (after application of block-matrix inversion this procedure requests at minimum inversion of two N_{BUS}-dimensional matrices [15]), it is not practical to construct state space matrix (A_{sys}) in (4.8) explicitly, since it will be dense. A second problem in the state space form (4.7) is that, again due to the inversion of $G_{\underline{V}}^{-1}$, for example, the influence of topology changes spreads to more entries, potentially varying them significantly.

For these reasons, it is more economical and practical to implement intended approaches for small-signal and DVS in *system (descriptor-based) form* as:

$$E \Delta \dot{X} = A' \Delta X \tag{4.11}$$

where:

$A', E \in \Re^{(n_x+n_{\underline{V}})\times(n_x+n_{\underline{V}})}$ – system (descriptor-form) matrices, where matrix $\boldsymbol{E}$ is singular;

$\boldsymbol{E} = \begin{bmatrix} \mathbf{1} & \mathbf{0} \\ \mathbf{0} & \mathbf{0} \end{bmatrix}$, where $\mathbf{1} \in \Re^{n_x \times n_x}$ is diagonal unity matrix, while $\mathbf{0}$ is zero matrix (with corresponding dimensions in matrix $\boldsymbol{E}$);

$A' = \begin{bmatrix} \boldsymbol{F}_x & \boldsymbol{F}_{\underline{V}} \\ \boldsymbol{G}_x & \boldsymbol{G}_{\underline{V}} \end{bmatrix}$, where matrix $\boldsymbol{G}_{\underline{V}}$ is nonsingular.

4.3 Eigenvalues and eigenvectors

The number of eigenvalues (possibly with repetitions; an eigenvalue is denoted by λ_i) is equal to the number of states ($i = 1, 2, \cdots, n$). They can be real or complex-conjugate pairs, as the system matrix is real. In most stability applications, it is of interest to determine whether the system equilibrium is at a bifurcation point, i.e., whether some eigenvalues of A_{sys} in (4.8) or A' in (4.11) have zero real part. For a stable power system, all the eigenvalues should have negative real parts (see Figure 4.2).

In practice, there are two characteristic cases of interest:[1]

1. One eigenvalue is $\lambda_i = 0$. This condition generally implies the occurrence of a *saddle-node bifurcation* [47–50].
2. A pair of complex eigenvalues as zero real part, i.e., $\lambda_{i,j} = \pm j\beta$. This condition generally implies the occurrence of a *HB* [51–54].

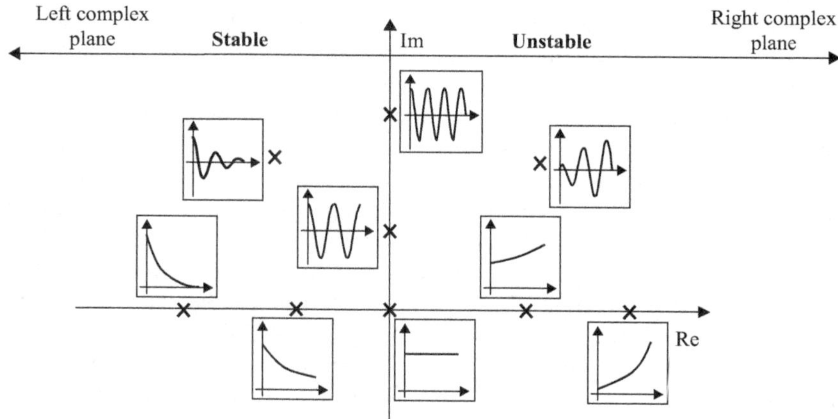

Figure 4.2 Positions of eigenvalues and typical time responses in complex plane

[1]Eigenvalue conditions for saddle-node and Hopf bifurcations are necessary but not sufficient. Proper transversality conditions that impose the dependence of the critical eigenvalues on system parameters complete the definitions of these bifurcation points [26, 44].

For each (i-th) eigenvalue (λ_i), there exists a right eigenvector ($\boldsymbol{v}_i$) and left eigenvector ($\boldsymbol{w}_i$) (with dimensions equal to the number of states (n), which satisfy equations of the form:

$$\boldsymbol{A}_{\text{sys}}\boldsymbol{v}_i = \lambda_i \boldsymbol{v}_i \tag{4.12}$$

$$\boldsymbol{w}_i^T \boldsymbol{A}_{\text{sys}} = \boldsymbol{w}_i^T \lambda_i \; i = 1, 2, \cdots, n \tag{4.13}$$

The right eigenvector indicates the relative magnitude of the participation of various states. The eigenvectors of a real mode are also real, and they are scaled so that the sum of the squares of the vector elements is equal to unity. Eigenvectors are not unique and may be multiplied by any scalar, and still be a valid eigenvector. However, the ratio between two elements is unique, provided that the eigenvalues are distinct.

Due to large size of the power system, it is often necessary to construct reduced-order models for dynamic stability studies by retaining only a few dominant modes. This requires a tool for identifying the state variables that have significant participation in a selected eigenvalue (e.g., an oscillation mode). It is natural to suggest that the significant state variables for an eigenvalue λ_i are those that correspond to large entries in the corresponding right eigenvector ($\boldsymbol{v}_i$). However, the entries in the eigenvector are dependent on the units of state variables that may be incommensurable (e.g., angle, velocity, flux).

Typically used eigenvalue solver is the Implicitly Restarted Arnoldi Method (IRAM) from ARPACK (implemented in Dynamic Security Assessment [DSA] Tools, PSS/E, and other commercial software packages), applied to the augmented linearized system matrices with variable complex shifts [46]. The complex shifts are automatically chosen to cover the required frequency-damping ratio ranges on the complex plane, where the computation starts at the higher frequency and damping ratio end of the specified search area [46].

Due to the excessive computation time necessary for obtaining the complete eigensolutions of large matrices, special methods have been developed which aim to compute a selected number of eigenvalues and the associated eigenvectors in studies of large-scale power systems. In one group of method, a limited number of eigenvalues associated with, for example, rotor angle modes are computed [55]. In several other methods eigenvalues associated with a small number of selected modes of interest are computed. These methods can handle large-scale power systems by taking advantage of network sparsity [5, 19, 56].

Participation factor analysis aids the identification of how each dynamic variable affects a given oscillation mode or eigenvalue. Participation factors are the best practical means to quickly identify the dominant states in an oscillation mode.

Specifically, for given a linear system:

$$\boldsymbol{x}(t_0) = \boldsymbol{x}_0; \quad \boldsymbol{x}(t_0) = \boldsymbol{x}_0 \tag{4.14}$$

then a participation factor is a sensitivity measure of an eigenvalue (λ_i) to a diagonal entry of the system matrix (A_{sys}):

$$p_{ki} = \frac{\partial \lambda_i}{\partial a_{sys,kk}} = \frac{w_{ki} v_{ik}}{\boldsymbol{w}_i^T \boldsymbol{v}_i} \tag{4.15}$$

where $a_{sys,kk}$ is the diagonal element of the system matrix A_{sys}, in the position kk.

In general, the participation factor (p_{ki}) is the measure of the influence of i-th state variable on k-th mode eigenvalue.

An eigenvector may be scaled by any value resulting in a new vector, which is also eigenvector. We can use this property to choose a scaling that simplifies the use of participation factors, for instance, choosing the eigenvectors such that their products is $\boldsymbol{w}_i^T \boldsymbol{v}_i = 1$; $i = 1, 2, \cdots, n$. In any case, since $\sum_{k=1}^{n} w_{ki} v_{ik} = \boldsymbol{w}_i^T \boldsymbol{v}_i$ it is:

$$\sum_{k=1}^{n} p_{ki} = 1. \tag{4.16}$$

This property is useful, since all participation factors are on a scale from zero to one. The eigenvectors corresponding to a complex eigenvalue will also have complex elements.

An alternative way for defining the participation factors follows from modal form of the original dynamic model (4.3) and (4.4) [4, 6]:

$$\dot{y} = \Lambda y \tag{4.17}$$

$$\boldsymbol{x} = \boldsymbol{T} \boldsymbol{y} \tag{4.18}$$

where:

$\Lambda = \boldsymbol{T}^{-1} \boldsymbol{A}_{sys} \boldsymbol{T}$ – diagonal matrix of eigenvalues;
$\boldsymbol{y} = \boldsymbol{T}^{-1} \boldsymbol{x}$ – new state vector in modal domain;
$\boldsymbol{T}$ – complex matrix whose columns are eigenvectors of $\boldsymbol{A}_{sys}$.

Each element y_i of vector $\boldsymbol{y}$ is termed as a *mode of the system*.

Let $\boldsymbol{Z} = \boldsymbol{T}^{-1}$, where $\boldsymbol{Z}$ is referred to as the "left eigenmatrix", and $\boldsymbol{T}$ as "right eigenmatrix". A participation factor is defined as:

$$p_{ik} = |t_{ik} z_{ki}| \tag{4.19}$$

where:

t_{ik} – the i-th element of the k-th column of matrix $\boldsymbol{T}$;
z_{ki} – the i-th element of the k-th row of matrix $\boldsymbol{Z}$.

Participation factors are nondimensional: the physical dimensions of the left eigenvector are the inverse of the physical dimensions of the corresponding right eigenvector. This makes it easier to use the participation vector to observe the relative importance of a particular state to a specific mode.

The eigenvalues of the linearized system matrix of coefficients ($\boldsymbol{A}_{\text{sys}}$), make it possible to determine the stability of the power system. The modes of time response correspond to these values.

The eigenvalue can be real or complex-conjugate pairs (see Figure 4.2), where:

- a real value relates to a mode with exponential time variation;
- the complex values, which appear in the form of complex-conjugate pairs, correspond to oscillatory modes.

The real part of the mode specifies the damping, while the imaginary part of each pair of complex eigenvalues gives the angular frequency of the oscillations, where:

- damped mode of oscillation exists if the real part is negative;
- undamped (divergent) mode of oscillation exists if the real part is positive.

For a complex-conjugate pair of eigenvalues ($\lambda = \alpha \pm j\beta$), the corresponding modes have the form:

$$x_1(t) = K_{c_1} e^{(\alpha + j\beta)t}; \quad x_2(t) = K_{c_2} e^{(\alpha - j\beta)t} \tag{4.20}$$

where the real part of the eigenvalue (α) determines the *damping* and the imaginary part (β) determines the *angular frequency* of the oscillations ($\omega = \beta$).

Then, the *frequency* of oscillations for a given eigenvalue pair ($\lambda = \alpha \pm j\beta$) is:

$$f = \frac{\text{Imag}\{\lambda\}}{2\pi} = \frac{\beta}{2\pi} \tag{4.21}$$

A good measure of an oscillation damping is the *damping ratio*, defined as:

$$\xi = \frac{-\alpha}{\sqrt{\alpha^2 + \beta^2}} (\text{positive for } \alpha < 0) \tag{4.22}$$

The power system electromechanical oscillations with damping ratios typically $\xi \geq 5\%$ are considered as satisfactory. But, this is not a hard rule. If the mode changes very little as the operating conditions are changed, a lower value of damping ratio (say $\xi = 3\%$) may be acceptable. However, when designing damping of the power system control, a damping ratio of at least 5% is an objective for the PSS, or the setting of the PSS in specific operating conditions.

Same characteristic examples are (1) with $\xi = 0$, oscillation does not decay; with $\xi = 5\%$, the oscillation decays to 10% for about 13 s; (2) with $\xi = 1\%$, the time of decay is about 6.5 s; and (3) with $\xi = 2\%$ about 3 s.

In power systems practice, damped electromechanical oscillations are not perceived as an issue; a problem appears only when the oscillations grow in magnitude with time. In general, the level of damping of $\xi = 5\%$ gives a small stability margin. The current design trend is to require the damping ratio (4.22) to be higher than 5%.

4.4 Necessary conditions for DVS equilibrium tracing

Although voltage stability involves dynamics, sometimes the power flow-based static analysis methods (*P–V* and *Q–V* analyses) are useful for fast (approximate) calculations and voltage stability assessment [2–8, 57].

The most important issue for voltage stability is to determine the risk of voltage instability in case of sudden disturbances. Another useful result is to predict (if possible) the voltage instability proximity indicators (e.g., [57]), including the effective means and/or measures that will prevent the occurrence of voltage collapse.

DVS is concerned with the ability of a power system to maintain acceptable voltages at all buses in the power system under normal conditions and after being subjected to a disturbance according to "($n - 1$)-concept" of security. The power system enters a state of *voltage instability* when a disturbance, an increase of load demand, or a change in system operating conditions, cause progressive and uncontrollable decline of voltage at any bus. The main factor causing instability is the ability of the power system to meet the reactive power demand. Conditions causing the voltage instability include the following typical situations:

- The power flow on the transmission corridor (tie-lines) is too high.
- The voltage/reactive power control resources are too far from the load centers.
- The source voltages (in generator buses) are too low.
- There is insufficient load reactive compensation level.

The method based on *P–V* curves detects the slower form of voltage instability, which occurs due to gradual increase in the power transfer between sending (source) and a receiving (sink) set of buses. The method is used also to size the reactive power compensation devices required at relevant buses to prevent voltage collapse (*Q–V* curves).

The *P–V* curve is a representation of voltage change as a result of increased active power transfer between two subsystems (source and sink). Tracing the *P–V* curves requires a parametric study involving a series of AC load flows or DAEs quasi steady-state solutions, that monitor the changes in one set of state and algebraic (usually bus voltages) variables with respect to another (usually branch active power flows). As power transfer is gradually increased (stepwise), voltage decreases at some buses on or near the transfer path.

A popular and robust technique to obtain full *P–V* (and *Q–V* curves) is the continuation method. This methodology basically consists of two power flow or DAEs quasi steady-state-based steps: the predictor and the corrector, as illustrated in Figure 4.3. In the predictor step, an estimate of the power flow (DAEs quasi steady state) solution for a load (P_R) increase (point 2 in Figure 4.3) is determined based on the starting solution (point 1) and an estimate of the changes in the power flow (DAEs quasi steady state) variables (e.g., state and bus complex voltages). This estimate may be computed using a linearization of the power flow (DAEs quasi steady state) equations, i.e., determining the "tangent vector" to the manifold of power flow (DAEs quasi steady state) solution.

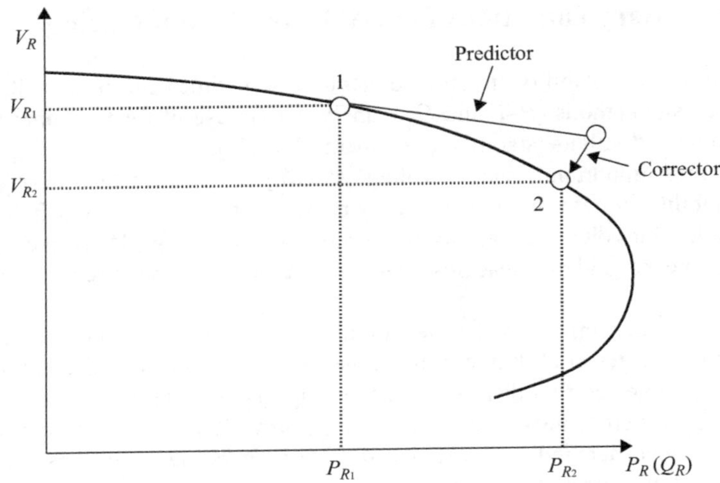

Figure 4.3 Predictor–corrector continuation power flow (DAEs quasi steady state) solution

The following optimization-based formulation can be used for the identification of critical conditions (determined by load/generation increments) for different types of bifurcation points described in Section 4.1 [26]:

$$\max_{\boldsymbol{X},\boldsymbol{v},\boldsymbol{v}',\zeta} \{\zeta\} \tag{4.23}$$

subject to:

1. Equilibrium constraints for differential equations from (4.3):

$$\boldsymbol{f}(\boldsymbol{X},\zeta) = \boldsymbol{0} \tag{4.24}$$

2. Equilibrium constraints for AEs from (4.4):

$$\boldsymbol{g}(\boldsymbol{X},\zeta) = \boldsymbol{0} \tag{4.25}$$

3. Necessary transversality conditions for different types of bifurcations (note that if all bifurcation types (described in Section 4.1) are analyzed simultaneously, particular equality constraints must be replaced with corresponding inequality constraints – for example, see (4.41) in Section 4.5):
   - LIBs, or boundary constraints for selected differential equations (***f***), AEs (***g***), state variables (***x***), algebraic variables (***V***), and other dependent functions and variables can be written in general form as:

$$\boldsymbol{h}^{\min} \leq \boldsymbol{h}(\boldsymbol{X},\zeta) \leq \boldsymbol{h}^{\max} \tag{4.26}$$

where constraints can be of any kind, for example, state variable limits, voltage limits, generator reactive power limits, transmission line flow (corridor) limits (defined by (4.46)–(4.49) in Section 4.5).

- HB is determined by the rightmost oscillatory eigenvalue $\lambda_{\max} = j\beta$:

$$\begin{bmatrix} \boldsymbol{F}_X \\ \boldsymbol{G}_X \end{bmatrix} \boldsymbol{v} = \begin{bmatrix} \mathrm{j}\beta \boldsymbol{v}_x \\ \boldsymbol{0} \end{bmatrix} \tag{4.27}$$

$$[\boldsymbol{F}_\zeta \quad \boldsymbol{G}_\zeta]\boldsymbol{v} \neq \boldsymbol{0} \tag{4.28}$$

$$\|\boldsymbol{v}\|_2 = 1 \tag{4.29}$$

where AEs for complex equation (4.27) are derived in Reference 35, while $\| \ \|_2$ indicates the Euclidean norm.

- SNB, as a special case of HB, where $\beta = 0$.
- SIB, determined by the eigenvalue closest-to-zero, $\lambda' = 0$:

$$\boldsymbol{J}_{\theta,V}\boldsymbol{v}' = \lambda'\boldsymbol{v}' \tag{4.30}$$

$$\boldsymbol{J}_\zeta \boldsymbol{v}' \neq \boldsymbol{0} \tag{4.31}$$

$$\|\boldsymbol{v}'\|_2 = \boldsymbol{1} \tag{4.32}$$

where:

ζ – load/generation change, included in equations for active and reactive bus injections (for constant power model), respectively (other load/generation models and single/distributed slack-node modeling are equivalent and do not introduce conceptual differences [2, 3, 23]):

$$\begin{aligned} &\left(1 + \zeta \frac{\Delta P_L}{\Delta P_G} K_{PGp}\right) P_{Gp_0} - (1 + \zeta K_{PLp}) P_{Lp_0} - V_p^2 G_{pp} \\ &\quad - V_p \sum_{q \in \alpha_p} \left[V_q (G_{pq} \cos\theta_{pq} + B_{pq} \sin\theta_{pq})\right] = 0 \end{aligned} \tag{4.33}$$

$$\begin{aligned} &Q_{Gp} - (1 + \zeta K_{QLp}) Q_{Lp_0} + V_p^2 B_{pp} \\ &\quad - V_p \sum_{q \in \alpha_p} \left[V_q (G_{pq} \sin\theta_{pq} - B_{pq} \cos\theta_{pq})\right] = 0 \end{aligned} \tag{4.34}$$

P_{Gp_0} – active power generation in p-th bus for basic case condition (where $\zeta = 0$);

P_{Lp_0}, Q_{Lp_0} – active and reactive loads in p-th bus for basic case condition (where $\zeta = 0$), respectively;

K_{PLp}, K_{QLp}, K_{PGp} – rate of active and reactive load changes as well as generator load pick-up factor in p-th bus (determined by automatic generation control, economic dispatch or other system operational practices), respectively[2];

$\Delta P_L = \zeta \sum_{p=1}^{N_{\text{BUS}}} K_{PLp} P_{Lp_0}$; $\Delta P_G = \zeta \sum_{p=1}^{N_{\text{BUS}}} K_{PGp} P_{Gp_0}$, where the balance equation for total active load/generation power increments must be satisfied [the balance equation for total reactive power change should be satisfied by power flow solution itself – see (4.37)]:

$$\sum_{p=1}^{N_{\text{BUS}}} \left(\zeta \frac{\Delta P_L}{\Delta P_G} K_{PGp} P_{Gp_0} - \zeta K_{PLp} P_{Lp_0} \right) + \Delta P_{\text{Losses}} = 0 \tag{4.35}$$

α_p – set of buses incident to p-th bus;
G_{pq}, B_{pq} – elements of $\underline{Y}_{\text{BUS}}$ matrix;

$$\boldsymbol{v} = \boldsymbol{v}_X = \begin{bmatrix} \boldsymbol{v}_x \\ \boldsymbol{v}_{\underline{V}} \end{bmatrix};$$

$\boldsymbol{v}_x$ – right eigenvector for state variables in DAEs model;
$\boldsymbol{v}_{\underline{V}}$ – extracted component of the right eigenvector corresponding to algebraic variables, which in general case can be calculated from right eigenvector component for state variables ($\boldsymbol{v}_x$) from (4.27) as ($\boldsymbol{G}_{\underline{V}}$ must be nonsingular matrix):

$$\boldsymbol{G}_{\underline{V}} \boldsymbol{v}_{\underline{V}} = -\boldsymbol{G}_x \boldsymbol{v}_x \tag{4.36}$$

$\boldsymbol{v}'$ – right eigenvector for power flow-based Jacobian matrix ($\boldsymbol{J}_{\theta,V}$);

$$\boldsymbol{F}_X = \left. \frac{\partial \boldsymbol{f}(\boldsymbol{X}, \zeta)}{\partial \boldsymbol{X}} \right|_{X_0,\zeta_0}; \quad \boldsymbol{G}_X = \left. \frac{\partial \boldsymbol{g}(\boldsymbol{X}, \zeta)}{\partial \boldsymbol{X}} \right|_{X_0,\zeta_0};$$

$$\boldsymbol{F}_\zeta = \left. \frac{\partial \boldsymbol{f}(\boldsymbol{X}, \zeta)}{\partial \zeta} \right|_{X_0,\zeta_0}; \quad \boldsymbol{G}_\zeta = \left. \frac{\partial \boldsymbol{g}(\boldsymbol{X}, \zeta)}{\partial \zeta} \right|_{X_0,\zeta_0};$$

$\boldsymbol{J}_{\theta,\underline{V}} = \left. \dfrac{\partial \boldsymbol{PQ}(\underline{\boldsymbol{V}}, \zeta)}{\partial \underline{\boldsymbol{V}}} \right|_{\underline{V}_0,\zeta_0} = \begin{bmatrix} \boldsymbol{J}_{P,\theta} & \boldsymbol{J}_{P,\underline{V}} \\ \boldsymbol{J}_{Q,\theta} & \boldsymbol{J}_{Q,\underline{V}} \end{bmatrix}$ – power flow-based Jacobian matrix [obtained by removing sensitivities of reactive power injections to voltages and sensitivities of loads to voltages in $\boldsymbol{G}_{\underline{V}}$ (4.5)], where: $\boldsymbol{J}_{P,\theta} = \left. \dfrac{\partial \boldsymbol{P}(\underline{\boldsymbol{V}}, \zeta)}{\partial \boldsymbol{\theta}} \right|_{\underline{V}_0,\zeta_0}$, $\boldsymbol{J}_{P,\underline{V}} = \left. \dfrac{\partial \boldsymbol{P}(\underline{\boldsymbol{V}}, \zeta)}{\partial \underline{\boldsymbol{V}}} \right|_{\underline{V}_0,\zeta_0}$; $\boldsymbol{J}_{Q,\theta} = \left. \dfrac{\partial \boldsymbol{Q}(\underline{\boldsymbol{V}}, \zeta)}{\partial \boldsymbol{\theta}} \right|_{\underline{V}_0,\zeta_0}$, $\boldsymbol{J}_{Q,\underline{V}} = \left. \frac{\partial \boldsymbol{Q}(\underline{\boldsymbol{V}}, \zeta)}{\partial \underline{\boldsymbol{V}}} \right|_{\underline{V}_0,\zeta_0}$ – $(N_{PQ} \times N_{PQ})$-, $(N_{PQ} \times N_{PV})$-, $(N_{PV} \times N_{PQ})$- and $(N_{PV} \times N_{PV})$-dimensional power flow-based Jacobian component matrices, respectively;

[2]Note that these values define the participation of loads/generations in source and sink areas for DVS tracing.

$\boldsymbol{PQ}(\underline{\boldsymbol{V}}, \zeta) = \begin{bmatrix} \boldsymbol{P}(\underline{\boldsymbol{V}}, \zeta) \\ \boldsymbol{Q}(\underline{\boldsymbol{V}}, \zeta) \end{bmatrix} = \begin{bmatrix} \boldsymbol{P}(\boldsymbol{\theta}, \boldsymbol{V}, \zeta) \\ \boldsymbol{Q}(\boldsymbol{\theta}, \boldsymbol{V}, \zeta) \end{bmatrix} = \begin{bmatrix} \boldsymbol{0} \\ \boldsymbol{0} \end{bmatrix}$ – sets of active (4.33) and reactive power bus injections (4.34) defined for all buses [slack-bus is excluded and defined by (4.35)];

P_{Losses}, Q_{Losses} – active and reactive power losses, respectively;

N_{PQ}, N_{PV} – total number of load (PQ) and generator (PV) buses, respectively.

The eigenvalue's real part provides a relative measure of proximity to system's instability, while the eigenvector, on the other hand, provides information related to the mechanism of loss of DVS [2, 3, 4, 6, 23]. The eigenvalues of DAEs model can be real and negative (or positive for unstable aperiodic condition) or complex conjugate (for oscillatory unstable condition). This means that right eigenvector for DAEs [$\boldsymbol{v}$ in (4.27)] can also have complex entries.

Furthermore, the existence of small eigenvalues of the power flow-derived Jacobian corresponding to the AEs determines the proximity of the system to the condition commonly labeled as *static voltage instability*. The power flow-based Jacobian matrix [$\boldsymbol{J}_{\theta,V}$ in (4.30)] is nonsymmetrical in general. This means that right eigenvector for AEs [$\boldsymbol{v}'$ in (4.30)] also may contain complex entries.[3] Once the minimum eigenvalue(s) and the corresponding left and right eigenvectors have been calculated, the participation factors can be used to identify the buses in the power system most affected by this condition [2–6, 23].

4.5 Optimization-based model for equilibrium tracing

The optimization problem from Section 4.4 is reformulated in the incremental form, where the initial guesses of right eigenvectors ($\boldsymbol{v}$ and $\boldsymbol{v}'$) are obtained from calculated rightmost (for DAEs) and closest-to-zero (for AEs) eigenvalues ($\lambda_{\max}$ and λ'), respectively.

Introducing a cut function (system's active and reactive power balance equations for perturbed conditions, $\zeta \neq 0$):

$$\gamma(\boldsymbol{X}, \zeta) = \gamma(\underline{\boldsymbol{V}}, \zeta) = \begin{cases} \zeta \sum_{p=1}^{N_{\text{BUS}}} \left(\dfrac{\Delta P_L}{\Delta P_G} K_{PGp} P_{Gp_0} - K_{PLp} P_{Lp_0} \right) + \Delta P_{\text{Losses}} = 0 \\ \sum_{p=1}^{N_{\text{BUS}}} (\Delta Q_{Gp} - \zeta K_{QLp} Q_{Lp_0}) + \Delta Q_{\text{Losses}} = 0 \end{cases} \tag{4.37}$$

the following linear programming model can be formulated:

$$\max_{\Delta \boldsymbol{X}, \Delta \boldsymbol{v}, \Delta \boldsymbol{v}', \Delta \zeta} \{\Delta \zeta\} \tag{4.38}$$

[3]For symmetric matrices, all eigenvalues are real; in addition, if the matrix is positive definite, then all its eigenvalues are (strictly) positive.

subject to constraints:

$$\boldsymbol{J}_{\text{aug}} \begin{bmatrix} \boldsymbol{X}_0 + \Delta \boldsymbol{X} \\ \boldsymbol{v}_0 + \Delta \boldsymbol{v} \\ \boldsymbol{v}'_0 + \Delta \boldsymbol{v}' \\ \zeta_0 + \Delta \zeta \end{bmatrix} = \begin{bmatrix} \mathbf{0} \\ \mathbf{0} \\ \mathbf{0} \\ \mathbf{0} \end{bmatrix} \tag{4.39}$$

$$\boldsymbol{h}^{\min} \le \boldsymbol{h}(\boldsymbol{X}_0, \zeta_0) + [\boldsymbol{H}_X \quad \boldsymbol{H}_\zeta] \begin{bmatrix} \Delta \boldsymbol{X} \\ \Delta \zeta \end{bmatrix} \le \boldsymbol{h}^{\max} \tag{4.40}$$

$$\mathrm{Re}\left\{ \begin{bmatrix} \boldsymbol{F}_X \\ \boldsymbol{G}_X \end{bmatrix} (\boldsymbol{v}_0 + \Delta \boldsymbol{v}) \right\} \le \begin{bmatrix} \mathbf{0} \\ \mathbf{0} \end{bmatrix} \tag{4.41}$$

$$[\boldsymbol{F}_\zeta \quad \boldsymbol{G}_\zeta](\boldsymbol{v}_0 + \Delta \boldsymbol{v}) < \text{ or } [\boldsymbol{F}_\zeta \quad \boldsymbol{G}_\zeta](\boldsymbol{v}_0 + \Delta \boldsymbol{v}) > \mathbf{0} \tag{4.42}$$

$$\mathrm{Re}\left\{ \left| \boldsymbol{J}_{\theta,\underline{V}}(\boldsymbol{v}'_0 + \Delta \boldsymbol{v}') \right| \right\} \ge \mathbf{0} \text{ and } \mathrm{Im}\left\{ \left| \boldsymbol{J}_{\theta,\underline{V}}(\boldsymbol{v}'_0 + \Delta \boldsymbol{v}') \right| \right\} \ge \mathbf{0} \tag{4.43}$$

$$\boldsymbol{J}_\zeta(\boldsymbol{v}'_0 + \Delta \boldsymbol{v}') < \mathbf{0} \text{ or } \boldsymbol{J}_\zeta(\boldsymbol{v}'_0 + \Delta \boldsymbol{v}') > \mathbf{0} \tag{4.44}$$

where:

$$\boldsymbol{J}_{\text{aug}} = \begin{bmatrix} \boldsymbol{F}_X & \boldsymbol{F}_v & \mathbf{0} & \mathbf{0} \\ \boldsymbol{G}_X & \boldsymbol{G}_v & \mathbf{0} & \mathbf{0} \\ \boldsymbol{J}_X & \mathbf{0} & \boldsymbol{J}_{v'} & \mathbf{0} \\ \nabla \boldsymbol{\gamma}_X & \mathbf{0} & \nabla \boldsymbol{\gamma}_{v'} & \nabla \boldsymbol{\gamma}_\zeta \end{bmatrix} = \begin{bmatrix} \boldsymbol{F}_x & \boldsymbol{F}_{\underline{V}} & \boldsymbol{F}_{v_x} & \boldsymbol{F}_{v_{\underline{\underline{V}}}} & \mathbf{0} & \mathbf{0} \\ \boldsymbol{G}_x & \boldsymbol{G}_{\underline{V}} & \boldsymbol{G}_{v_x} & \boldsymbol{G}_{v_{\underline{\underline{V}}}} & \mathbf{0} & \mathbf{0} \\ \mathbf{0} & \boldsymbol{J}_{\underline{V}} & \mathbf{0} & \mathbf{0} & \boldsymbol{J}_{v'} & \mathbf{0} \\ \mathbf{0} & \nabla \boldsymbol{\gamma}_{\underline{\underline{V}}} & \mathbf{0} & \mathbf{0} & \nabla \boldsymbol{\gamma}_{v'} & \nabla \boldsymbol{\gamma}_\zeta \end{bmatrix} \tag{4.45}$$

$$\boldsymbol{H}_X = \left. \frac{\partial \boldsymbol{h}(\boldsymbol{X}, \zeta)}{\partial \boldsymbol{X}} \right|_{\boldsymbol{X}_0, \zeta_0} \quad \boldsymbol{H}_\xi = \left. \frac{\partial \boldsymbol{h}(\boldsymbol{X}, \zeta)}{\partial \zeta} \right|_{\boldsymbol{X}_0, \zeta_0}$$

$$\Delta P_{\text{Losses}} = \left. \frac{\partial P_{\text{Losses}}(\underline{V}, \zeta)}{\partial \underline{V}} \right|_{\underline{V}_0, \zeta_0} \Delta \underline{V} \Delta Q_{\text{Losses}} = \left. \frac{\partial Q_{\text{Losses}}(\underline{V}, \zeta)}{\partial \underline{V}} \right|_{\underline{V}_0, \zeta_0} \Delta \underline{V};$$

$$\Delta Q_{Gp} = \left. \frac{\partial Q_{Gp}(\underline{V}, \zeta)}{\partial \underline{V}} \right|_{\underline{V}_0, \zeta_0} \Delta \underline{V}$$

In the case when analytical derivations for calculation of gradient $\nabla \boldsymbol{\gamma}_{v'}$ and similar gradients in (4.45) are unavailable, a numerical alternative is finite differencing [26, 35].

Conditional constraint that at least one of the constraints (4.43) or (4.44) needs to be satisfied is captured by additional binary variables and the corresponding additional constraint (two conditional constraints introduce one additional constraint) [58].

For general form of boundary constraints (4.40), the following particular types of constraints (denoted with upper index "c") are considered here: (1) state variables (e.g., generator field voltages, exciter output voltages, exciter saturation limits,), (2) generator reactive power injections, (3) bus voltage magnitudes, and (4) branch (and/or corridor) power flows, respectively:

$$\boldsymbol{x}^{c,\min} \leq \boldsymbol{x}^{c} = \boldsymbol{x}^{c}(\underline{\boldsymbol{V}}_0, \zeta_0) + \Delta\boldsymbol{x}^{c} \leq \boldsymbol{x}^{c,\max} \tag{4.46}$$

$$\boldsymbol{Q}_G^{c,\min} \leq \boldsymbol{Q}_G^{c} = \boldsymbol{Q}_G^{c}(\underline{\boldsymbol{V}}_0, \zeta_0) + \nabla\boldsymbol{Q}_{G,\underline{\boldsymbol{V}}}^{c}\Delta\underline{\boldsymbol{V}} \leq \boldsymbol{Q}_G^{c,\max} \tag{4.47}$$

$$\boldsymbol{V}^{c,\min} \leq \boldsymbol{V}^{c} = \boldsymbol{V}_0^{c}(\zeta_0) + \Delta\boldsymbol{V}^{c} \leq \boldsymbol{V}^{c,\max} \tag{4.48}$$

$$\boldsymbol{P}_b^{c,\min} \leq \boldsymbol{P}_b^{c} = \boldsymbol{P}_b^{c}(\underline{\boldsymbol{V}}_0, \zeta_0) + \nabla\boldsymbol{P}_{b,\underline{\boldsymbol{V}}}^{c}\Delta\underline{\boldsymbol{V}} \leq \boldsymbol{P}_b^{c,\max} \tag{4.49}$$

where:

$\boldsymbol{x}^c$ – vector of constrained state variables (calculated from power flow solution ($\underline{\boldsymbol{V}}$) and steady-state DAEs);
$\boldsymbol{Q}_G^c$ – vector of constrained reactive power outputs (in PV buses), calculated from the reactive power bus injection (4.34);
$\boldsymbol{V}^c$ – vector of constrained bus voltage magnitudes (typically in PQ buses);
$\boldsymbol{P}_b^c$ – vector of branch (and/or transmission corridor) active power flows;
$\nabla\boldsymbol{Q}_{G,\underline{\boldsymbol{V}}}^{c} = \left.\dfrac{\partial\boldsymbol{Q}_G^{c}(\underline{\boldsymbol{V}},\zeta)}{\partial\underline{\boldsymbol{V}}}\right|_{\underline{\boldsymbol{V}}_0,\zeta_0}$ $\nabla\boldsymbol{P}_{b,\underline{\boldsymbol{V}}}^{c} = \left.\dfrac{\partial\boldsymbol{P}_b^{c}(\underline{\boldsymbol{V}},\zeta)}{\partial\underline{\boldsymbol{V}}}\right|_{\underline{\boldsymbol{V}}_0,\zeta_0}$;

4.6 Iterative algorithms for DVS tracing

The optimization problem for DVS tracing specified in Section 4.5 is challenging, because the eigenvector entries vary drastically close to the bifurcation point(s). Since this formulation requires only the critical (rightmost and closest-to-zero) eigenvalues and a constrained power flow solution, it is amenable to alternative tractable and scalable algorithms for DVS tracing.

4.6.1 Predictor–corrector-based algorithm

Details of the predictor–corrector-based algorithm are described by the following steps [26]:

1. PREDICTOR

1a. Calculation of critical (rightmost and closest-to-zero) eigenvalues

For current equilibrium point ($\boldsymbol{X}_0 = \begin{bmatrix}\boldsymbol{x}_0^T & \underline{\boldsymbol{V}}_0^T\end{bmatrix}^T$), satisfying $\boldsymbol{f}(\boldsymbol{X}_0, \zeta_0) = \boldsymbol{0}$, $\boldsymbol{g}(\boldsymbol{X}_0, \zeta_0) = \boldsymbol{0}$ and $\boldsymbol{PQ}(\boldsymbol{X}_0, \zeta_0) = \boldsymbol{0}$, calculate the critical eigenvalue with largest real part (rightmost) for DAEs model ($\lambda_{\max}$) and eigenvalue closest-to-zero for AEs (power flow-based) model (λ') by algorithm proposed in Reference 56.

1b. Calculation of right eigenvectors
Calculation of corresponding right eigenvectors ($\boldsymbol{v}$ and $\boldsymbol{v}'$) for critical (rightmost and closest-to-zero) eigenvalues ($\lambda_{\max}$ and λ'), respectively, as:

$$\begin{bmatrix} \boldsymbol{F}_x & \boldsymbol{F}_{\underline{V}} \\ \boldsymbol{G}_x & \boldsymbol{G}_{\underline{V}} \end{bmatrix} \begin{bmatrix} \boldsymbol{v}_{x0} \\ \boldsymbol{v}_{\underline{V}0} \end{bmatrix} = \lambda_{\max} \begin{bmatrix} \mathbf{1} & \mathbf{0} \\ \mathbf{0} & \mathbf{0} \end{bmatrix} \begin{bmatrix} \boldsymbol{v}_{x0} \\ \boldsymbol{v}_{\underline{V}0} \end{bmatrix} \tag{4.50}$$

$$\boldsymbol{J}_{\theta,V} \boldsymbol{v}'_0 = \lambda' \boldsymbol{v}'_0 \tag{4.51}$$

1c. Calculation of increments
Solve the optimization problem (4.38)–(4.44).
If $\Delta\zeta = 0$, one from analyzed types of bifurcation points (SNB, HB, SIB, or LIB) is reached (bifurcation manifold).

1d. Prediction
Predicted state ($\boldsymbol{x}^{(k+1)}$) and algebraic ($\underline{\boldsymbol{V}}^{(k+1)}$) variables as well as load/generation change ($\zeta^{(k+1)}$), respectively, are:

$$\boldsymbol{X}^{(k+1)} = \begin{bmatrix} \boldsymbol{x}^{(k+1)} \\ \underline{\boldsymbol{V}}^{(k+1)} \end{bmatrix} = \boldsymbol{X}^{(k)} + \Delta\boldsymbol{X} \tag{4.52}$$

$$\zeta^{(k+1)} = \zeta^{(k)} + \Delta\zeta \tag{4.53}$$

2. LOCAL PARAMETERIZATION

Calculation of DAEs and AEs (power flow-based) increments in two successive iterations:

$$\Delta\boldsymbol{z}(\boldsymbol{X}^{(k+1)}, \zeta^{(k+1)}) = \begin{bmatrix} \boldsymbol{f}(\boldsymbol{X}^{(k+1)}, \zeta^{(k+1)}) - \boldsymbol{f}(\boldsymbol{X}^{(k)}, \zeta^{(k)}) \\ \boldsymbol{g}(\boldsymbol{X}^{(k+1)}, \zeta^{(k+1)}) - \boldsymbol{g}(\boldsymbol{X}^{(k)}, \zeta^{(k)}) \\ \boldsymbol{PQ}(\underline{\boldsymbol{V}}^{(k+1)}, \zeta^{(k+1)}) - \boldsymbol{PQ}(\underline{\boldsymbol{V}}^{(k+1)}, \zeta^{(k)}) \\ \boldsymbol{\gamma}(\underline{\boldsymbol{V}}^{(k+1)}, \zeta^{(k+1)}) - \boldsymbol{\gamma}(\underline{\boldsymbol{V}}^{(k+1)}, \zeta^{(k)}) \end{bmatrix} \tag{4.54}$$

3. CORRECTOR: Adjust the initially predicted solution

The Newton–Raphson-based method is employed to do the boundary corrections as:

$$\boldsymbol{J}'_{\text{aug}} \begin{bmatrix} \boldsymbol{X}^{*(k+1)} \\ \zeta^{*(k+1)} \end{bmatrix} = -\Delta\boldsymbol{z}(\boldsymbol{X}^{(k+1)}, \zeta^{(k+1)}) \tag{4.55}$$

where $\boldsymbol{J}'_{\text{aug}}$ correspond to the components $\boldsymbol{F_X}$, $\boldsymbol{G_X}$, $\boldsymbol{J_X}$, $\nabla\boldsymbol{\gamma_X}$ and $\nabla\boldsymbol{\gamma}_\zeta$ in $\boldsymbol{J}_{\text{aug}}$ (4.45).

4.6.2 Interval bisection-based algorithm

The interval bisection-based algorithm for DVS tracing is described in the following steps:

1. Define the interval for bisection [$\zeta_a = 0$; ζ_b]

   For $\zeta_a = 0$ is supposed that the power system is stable (lower/upper bounds for $\boldsymbol{h}(\boldsymbol{X}, \zeta)$ are satisfied (constrained power flow converges), $\mathrm{Re}\{\lambda_{\max}\} < 0$ and $\lambda', \lambda'' \neq 0$).
2. Solve Constrained Power Flow

   For actual (for first iteration in point *b* and in other iterations in point *c*) loading margin (ζ) solve constrained power flow solution specified by equality (4.25) and inequality constraints (4.46)–(4.49).
3. Calculation of critical eigenvalues

   For loading margin (ζ), calculate the critical eigenvalues [rightmost ($\lambda_{\max}$) and closest-to-zero (λ' and λ'', respectively)] [56].
4. Check the bifurcation type

   For actual loading margin (ζ), check the following DVS conditions:
   - For divergent constrained power flow ([typically where violated any type of constraints $\boldsymbol{h}^{\min} \leq \boldsymbol{h}(\boldsymbol{X}, \zeta) \leq \boldsymbol{h}^{\max}$, or (4.46)–(4.49)), we have LIB and the maximum loading margin is $\zeta_{\max} = \zeta_{a(b)}$.
   - For $\mathrm{Re}\{\lambda_{\max}\} > 0$, the power system is unstable; if $\lambda_{\max} = j\beta$, we have HB and the maximum loading margin is $\zeta_{\max} = \zeta_{a(b)}$.
   - For $\lambda_{\max} \approx 0$, we have SNB and the maximum loading margin is $\zeta_{\max} = \zeta_{a(b)}$.
   - For convergent constrained power flow and if $\mathrm{Re}\{\lambda'\}$ at two interval bounds (*a* and *b*) are with opposite signs, we have $SIB_{P\theta,QV}$; if $\lambda' \approx 0$, the maximum loading margin is $\zeta_{\max} = \zeta_{a(b)}$.
   - For convergent constrained power flow and if $\mathrm{Re}\{\lambda''\}$ at two interval bounds (*a* and *b*) are with opposite signs, we have SIB_{QV}; if $\lambda'' \approx 0$, the maximum loading margin is $\zeta_{\max} = \zeta_{a(b)}$.
5. Bisection of the loading interval

   Divide the loading interval in two subintervals by computing the midpoint $\zeta_c = (\zeta_a + \zeta_b)/2$ and check bifurcation type for ζ_c (Steps 3 and 4) at that point. Select the subinterval that is a bracket as a new interval to be used in the next step ($\zeta_a = \zeta_c$ or $\zeta_b = \zeta_c$). This procedure implies that the interval containing a bifurcation point is halved at each iteration step.

   The iterative process is continued until the loading interval is sufficiently small and/or one of necessary bifurcation conditions from Step 4 is satisfied.

4.6.3 Q–V *Sensitivity*

The eigenvalue associated with a mode of bus reactive power-voltage (*Q*–*V*) variation (λ'') can provide a relative measure of proximity to voltage instability and thus complement the previously introduced eigenvalues $\lambda_{\max}$ (4.50) and λ' (4.51). From the power flow Jacobian matrix [$\boldsymbol{J}_{\theta,V}$ in (4.30)], we have $\Delta Q/\Delta V = \boldsymbol{J}_R = \boldsymbol{J}_{Q,V} - \boldsymbol{J}_{Q,\theta}\boldsymbol{J}_{P,\theta}^{-1}\boldsymbol{J}_{P,V}$, where the matrix $\boldsymbol{J}_R$ represents the linearized relationship between the incremental changes in p-th bus voltage (ΔV_p) and reactive power injection (ΔQ_p). Note that inversion $\boldsymbol{J}_{P,\theta}^{-1}$ would be too slow [15], and eigenvalue λ'' is calculated from descriptor-based form, similarly as in (4.6), as [44]:

$$\begin{bmatrix} \mathbf{0} & \mathbf{0} \\ \mathbf{0} & \mathbf{1} \end{bmatrix} \begin{bmatrix} \Delta \boldsymbol{P}/\Delta \boldsymbol{\theta} \\ \Delta \boldsymbol{Q}/\Delta \boldsymbol{V} \end{bmatrix} = \begin{bmatrix} \boldsymbol{J}_{P,\theta} & \boldsymbol{J}_{P,V} \\ \boldsymbol{J}_{Q,\theta} & \boldsymbol{J}_{Q,V} \end{bmatrix} \tag{4.56}$$

The magnitude of the eigenvalue λ'' provides a relative measure of proximity to the Q–V instability (SIB$_{QV}$ bifurcation). Then, the participation factors can be used effectively to find out the weakest buses in the power system.

4.7 Dynamic analysis with topology control (TC) optimization

With increasing integration of intermittent wind and solar generation, and with market-driven developments, unplanned changes in power flows have increased in both frequency and magnitude. Dynamic transmission TC is one of several emerging control technologies that are able to respond to insecure power system operation [15, 59]. The TC actions are typically followed by generators' economic redispatch. TC offers significant value to the system operator through more efficient use of existing transmission resources. Although TC has been a topic of ongoing interest in the research community for a number of years [60, 61], the scalable and practical solutions have not been presented. Neither, however, has there been a significant demand by the power industry as traditional sources of variability and uncertainty could be managed without dynamic methods. In the past, operations have been supported by a robust transmission grid and by well-defined patterns of power flows. However, with increasing integration of wind and solar generation and with market-driven developments, unplanned changes in power flows have increased in both frequency and magnitude. Current approaches used to manage the limited transmission capacity rely primarily on varying generation with limited changes of transmission topology. Reconfiguration of lines is limited to outage contingency planning and is implemented infrequently for reliability purposes, while precontingency grid topology is typically not modified for economic dispatch. Therefore, since power flows are dictated by Kirchhoff's laws, generation redispatch (replacing low-cost generation limited by network congestion with higher-cost alternatives at appropriate locations) is the only tool available to operators to avoid transmission overloads.

Potential economic benefits of optimized TC have been found to be significant in a number of very different systems, with production cost savings of several percentage points over the solution when the transmission topology is fixed, and even with the inclusion of security constraints [60]. A related concern is that reactive power, small-signal, transient, and voltage stability limits may become critical under congestion conditions, and line status changes may both improve and worsen reliability [62]. Voltage instability sometimes leads to the blackout of a large portion or even the whole interconnected power system.

4.7.1 DVS Assessment

The TC optimization is a technique used by transmission system operators to manage switches during periods of constraint in order to allow for greater energy flows on the system and to reduce overall system costs [59]. The TC optimization simultaneously optimizes the grid configuration and generation (sometimes also demand) resources, while providing a list of operator switching actions for in-field implementation of the optimal solution.

In-field implementation of the TC switching actions sometimes can generate unacceptable transient oscillations and smaller load margins for DVS. The power system needs to stay in the feasible region during the transient and steady-state conditions for TC actions. In Figure 4.4, a flowchart illustrating an algorithm for transient stability and DVS assessment with application of the TC actions is shown.

Note that short-term dynamic model (4.3), (4.4) was used for transient stability assessment after TC switching actions (TC branch switching list and generation redispatch list – long-term dynamics is neglected), while the dynamic voltage tracing is performed on the quasi-state model, as shown in (4.24).

4.7.2 Small-signal stability analysis

This section explores the influence of TC changes and generation redispatch on fast small-signal stability (SSS) assessment. The SSS problem of a large power system is usually one of insufficient damping for system oscillations, and its analysis is based on linearized system dynamic equations using modal (eigenvalue) analysis techniques [2–8]. After topology changes, a fast algorithm for update of system matrices (in descriptor form) without matrix inversions is used. To additionally reduce the computational time, only critical eigenvalues in base case operating condition (e.g., rightmost or those in given damping ratio and frequency ranges) are used for calculation of closest eigenvalue perturbations for TC changes. Using the simplified algorithm, all TC changes are prefiltered to detect critical ones from the SSS constraint standpoint. Only critical TC changes are reevaluated by full dynamic model for perturbation of output active power for participating generators in critical eigenvalues. The sensitivities of eigenvalues to generation redispatch are calculated using eigenvalue perturbations from base case operating condition to TC cases. These sensitivities determine the SSS constraints for optimization of eigenvalue placement, or for global TC optimization. These steps enable a tractable algorithm for calculations in large-scale power systems. We developed MATLAB® software that is integrated with a professional software package for power flow, small-signal and transient analysis (DSA Tools [46]) enabling the exchange of system matrices and/or generation redispatch between them. Our procedure can also be used to study dynamic aspects of line or other equipment maintenance. Our approach relies, of course, on a dynamic model of the power system considered, further emphasizing the need for a complete and verified database.

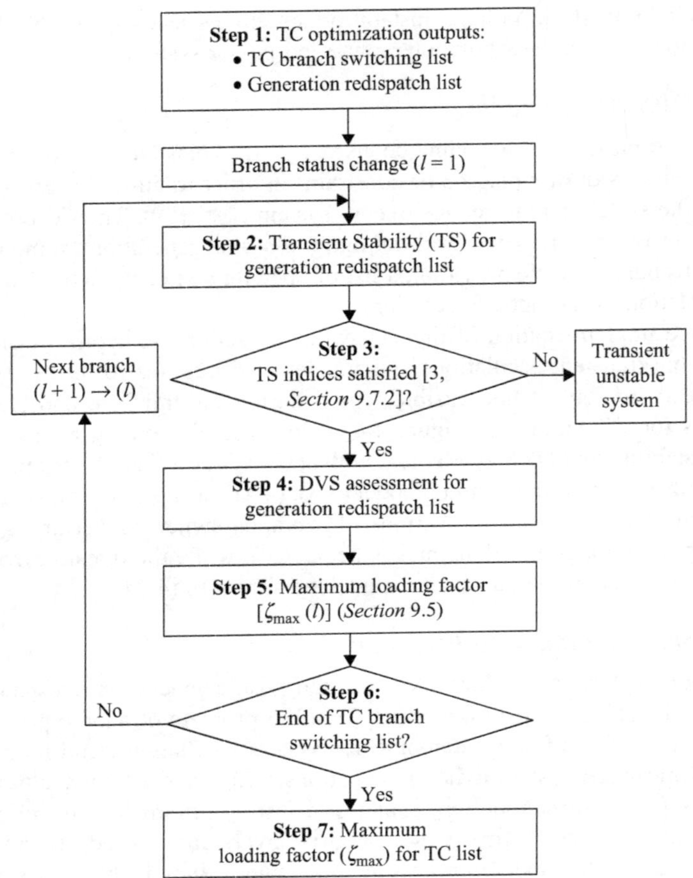

Figure 4.4 Flowchart of transient analysis and DVS assessment with TC optimization actions (Fig. 1 1 2016 IEEE. Reprinted with permission from Reference 26)

From derived sensitivity in (4.6), $\boldsymbol{G}_{\underline{V}} = \boldsymbol{G'}_{\underline{V}} - \underline{\boldsymbol{Y}'}_{\text{BUS}}$ is composed from component real (denoted by index r) and imaginary (denoted by index i) part submatrices as:

$$\boldsymbol{G}_{\underline{V}} = \begin{bmatrix} \boldsymbol{G'}_{\underline{V},rr} - \underline{\boldsymbol{Y}'}_{\text{BUS},rr} & \boldsymbol{G'}_{\underline{V},ri} - \underline{\boldsymbol{Y}'}_{\text{BUS},ri} \\ \boldsymbol{G'}_{\underline{V},ir} - \underline{\boldsymbol{Y}'}_{\text{BUS},ir} & \boldsymbol{G'}_{\underline{V},ii} - \underline{\boldsymbol{Y}'}_{\text{BUS},ii} \end{bmatrix} \tag{4.57}$$

Change in transmission branch on–off status dominantly influences to the power flow-based bus-branch admittance matrix ($\Delta\underline{\boldsymbol{Y}'}_{\text{BUS}} \approx \Delta\underline{\boldsymbol{Y}}_{\text{BUS}}$), with an assumption that network's voltage profile is not changed significantly [this means that matrices $\boldsymbol{F}_{\boldsymbol{x}}$, $\boldsymbol{F}_{\underline{V}}$, $\boldsymbol{G}_{\boldsymbol{x}}$, and $\boldsymbol{G'}_{\underline{V}}$ in (4.5) are the same before and after the change in transmission branch status]. For (k)-th branch (between s-th and t-th buses), status change (line opening) in TC process, matrix $\underline{Y}_{\mathrm{BUS}}$ is modified by [63]:

$$\Delta\underline{Y}'_{\mathrm{BUS},z} \approx \Delta\underline{Y}_{\mathrm{BUS},z} = \underline{Y}^{(k)}_{\mathrm{BUS},z} - \underline{Y}^{(0)}_{\mathrm{BUS},z} = -\underline{Y}^{b}_{st,z}\boldsymbol{e}_k\boldsymbol{e}_k^T \tag{4.58}$$

where:

$\underline{Y}^{b}_{\mathrm{st},z}$ – admittance of the branch between s-th and t-th buses for Base Case (no topology change) operating condition and $z = rr,\ ri,\ ir,\ ii$;
$\boldsymbol{e}_k = [0 \quad \cdots \quad 1 \quad \cdots \quad -1 \quad \cdots \quad 0]^T$, where for s-th position is $e_{k,s} = 1$ and for t-th position is $e_{k,t} = -1$.

The (k)-th topology change influences to the matrix $\boldsymbol{G}_{\underline{V}}$ (4.5) as:

$$\begin{aligned}\boldsymbol{G}^{(k)}_{\underline{V},z} &= \boldsymbol{G}'^{(0)}_{\underline{V},z} - \underline{Y}^{(k)}_{\mathrm{BUS},z} = \boldsymbol{G}'^{(0)}_{\underline{V},z} - (\underline{Y}^{(0)}_{\mathrm{BUS},z} + \Delta\underline{Y}_{\mathrm{BUS},z}) \\ &= \boldsymbol{G}'^{(0)}_{\underline{V},z} - (\underline{Y}^{(0)}_{\mathrm{BUS},z} - \underline{Y}^{b}_{st,z}\boldsymbol{e}_k\boldsymbol{e}_k^T)\end{aligned} \tag{4.59}$$

Submatrices $\boldsymbol{F}_x$, $\boldsymbol{F}_{\underline{V}}$, $\boldsymbol{G}_x$, and $\boldsymbol{G}_{\underline{V}}$ are extremely sparse, so sparse matrix techniques can be efficiently employed for the storage and evaluation of the particular elements.

In cases of significant topology changes (heavily loaded lines, large transformers, etc.), the assumption about unchanged bus voltages typically leads to the large errors in the calculation of eigenvalue movements. For fast calculation of perturbations in voltage-dependent power flow-based bus-branch admittance matrix ($\underline{Y}_{\mathrm{BUS}}$) after topology changes, please consult References 64 and 65. In cases where values of the system dynamic matrices ($\boldsymbol{F}_x$, $\boldsymbol{F}_{\underline{V}}$, $\boldsymbol{G}_x$, and $\boldsymbol{G}'_{\underline{V}}$) are changed significantly with voltage perturbations, and their recalculation using the full dynamic model is needed. The recalculation of the system matrices by full dynamic model is also required in case of voltage-dependent and/or dynamic loads. The computation time for updating the system matrices is not critical for the performance of this approach.

Damping ratio of eigenvalue $\lambda_i = \sigma_i + j\omega_i$ [$i = 1, 2, \cdots, n_c$, where n_c is total number of critical (e.g., rightmost) eigenvalues], is given by (4.22) and rewritten as:

$$\xi_i = -\frac{\sigma_i}{\sqrt{\sigma_i^2 + \omega_i^2}} \tag{4.60}$$

where σ_i and ω_i are damping and frequency of analyzed i-th complex-conjugate eigenvalue, respectively.

Based on simulations presented in References 15 and 64, the following assumption for (k)-th branch status change (branch opening) is introduced:

$$\Delta\sigma^{(k)}_{ij} = K^{(k)}_{\sigma,ij}\Delta P^{(k)}_{Gj};\ \Delta\omega^{(k)}_{ij} \approx 0 \tag{4.61}$$

where:

$$K_{\sigma,ij}^{(k)} = \frac{\Delta\sigma_{ij}^{(k)}}{\Delta P_{Gj}^{(k),sp}} = \frac{\sigma_{ij}^{(k),sp} - \sigma_i^{(k)}}{P_{Gj}^{(k),sp} - P_{Gj}^{(0)}} \tag{4.62}$$

and:

$\sigma_i^{(k)}$, $\sigma_i^{(k),sp}$ – damping of i-th eigenvalue before and after specified j-th generator's output active power perturbation for (k)-th topology change ($j \in N_{Gi}^{(k)}$), respectively;

$\Delta P_{Gj}^{(k),sp}$ – the j-th generator's output active power perturbation (by the user-specified value; for example, $\Delta P_{Gj}^{(k),sp} = P_{Gj}^{(k),sp} - P_{Gj}^{(0)} = 5\% P_{Gj}^{(0)}; j \in N_{Gi}^{(k)}$);

$N_{Gi}^{(k)}$ – set of generators that dominantly influenced by electromechanical state variables to the i-th eigenvalue for (k)-th topology change.

After a single (j-th) generator's output active power perturbation (ΔP_{Gj}), from (4.61) we have:

$$\xi_{ij}^{(k)} \approx -\frac{\sigma_i^{(0)} + K_{\sigma,ij}^{(k)}\Delta P_{Gj}^{(k)}}{\sqrt{\left(\sigma_i^{(0)} + K_{\sigma,ij}^{(k)}\Delta P_{Gj}^{(k)}\right)^2 + \left(\omega_i^{(0)}\right)^2}} \tag{4.63}$$

From (4.61), the increment of damping due to the perturbation of all generator's output active powers is:

$$\Delta\sigma_i^{(k)} = \sum_{j\in N_{Gi}^{(k)}} \Delta\sigma_{ij}^{(k)} = \sum_{j\in N_{Gi}^{(k)}} K_{\sigma,ij}^{(k)}\Delta P_{Gj} \geq \Delta\sigma'^{(k)}_i \tag{4.64}$$

where $\Delta\sigma'^{(k)}_i = \sigma_i^{(k),\text{req}} - \Delta\sigma_i^{(k)}$ and $\sigma_i^{(k),\text{req}}$ is minimum requested damping of i-th eigenvalue and (k)-th topology change for requested damping ratio (e.g., $\xi^{\text{req}} = 5\%$):

$$\sigma_i^{(k),\text{req}} = -\frac{\xi^{\text{req}}}{\sqrt{1 - (\xi^{\text{req}})^2}}\omega_i^{(k)} \tag{4.65}$$

while $N_{Gi}^{(k)}$ is set of generators that participate in i-th eigenvalue [for (k)-th TC change].

In matrix notation, from (4.64) for (k)-th topology change and all critical eigenvalues ($i = 1, 2, \cdots, n_c$) is:

$$\Delta\sigma^{(k)} = \boldsymbol{K}_\sigma^{(k)}\Delta\boldsymbol{P}_G^{(k)} \geq \Delta\sigma'^{(k)} \tag{4.66}$$

where $\boldsymbol{K}_\sigma^{(k)}$ is sensitivity matrix for (k)-th topology change, with elements defined in (4.62). This result is, of course, local and thus limited to small perturbations, but it works surprisingly well in the cases we encountered. It is assumed that the system operates as a single interconnection for economic reasons, thus excluding the possibility of islanding as an operating mode.

4.7.3 Optimization of eigenvalues movement

The problem of eigenvalues movement with TC changes can be posed as a (local) quadratic optimization, where the objective function is to minimize generator output active power perturbations (over $k = 1, 2, \cdots, K_{TC}$, where K_{TC} is number of analyzed TC changes):

$$\min_{\Delta \boldsymbol{P}_{Gj}^{(k)}} \left\{ \sum_j (\Delta P_{Gj}^{(k)})^2 \right\} \tag{4.67}$$

subject to:

$$\sum_j \Delta P_{Gj}^{(k)} = 0 \tag{4.68}$$

$$\boldsymbol{K}_{\sigma}^{(k)} \Delta \boldsymbol{P}_G^{(k)} \geq \Delta \sigma'^{(k)} \tag{4.69}$$

$$\boldsymbol{P}_G^{\min} - \boldsymbol{P}_G^{(0)} \leq \Delta \boldsymbol{P}_G \leq \boldsymbol{P}_G^{\max} - \boldsymbol{P}_G^{(0)} \tag{4.70}$$

A tractable and scalable algorithm for fast assessment of SSS, eigenvalue sensitivities to simultaneous TC changes and generations redispatch, as well as optimization of eigenvalues movement to satisfy SSS constraints, can be summarized in the following steps (Figure 4.5):

Step 1: Calculation of critical eigenvalues for base case operating condition ($\boldsymbol{\lambda}^{(0)}$). Discussion about calculation of critical eigenvalues is provided in Section 4.8.2.

Step 2: For selected (k)-th topology change, fast update of system matrices (in descriptor form) by simplified algorithm (Section 4.7.2).

Step 3: Calculation of closest eigenvalues to the critical eigenvalues in base case operating condition ($\boldsymbol{\lambda}^{(0)}$) for ($k$)-th topology change ($\boldsymbol{\lambda}^{(k)}$).

Step 4: Check the validation criteria for simplified model, where following single or all criteria can be used:

1. existence of low-damped eigenvalues:

$$\xi_i^{(k)} \leq \xi^{\text{req}}; \quad i = 1, 2, \cdots, n_c \tag{4.71}$$

2. maximum eigenvalue perturbation:

$$\max_i \left(\frac{\left| \left| \lambda_i^{(0)} \right| - \left| \lambda_i^{(k)} \right| \right|}{\left| \lambda_i^{(0)} \right|} \right) \leq \varepsilon_1; \quad i = 1, 2, \cdots, n_c \tag{4.72}$$

3. average eigenvalue perturbation:

$$\frac{1}{n_c} \sum_i \frac{\left| \left| \lambda_i^{(0)} \right| - \left| \lambda_i^{(k)} \right| \right|}{\left| \lambda_i^{(0)} \right|} \leq \varepsilon_2 \tag{4.73}$$

where ε_1 are ε_2 tolerance criteria.

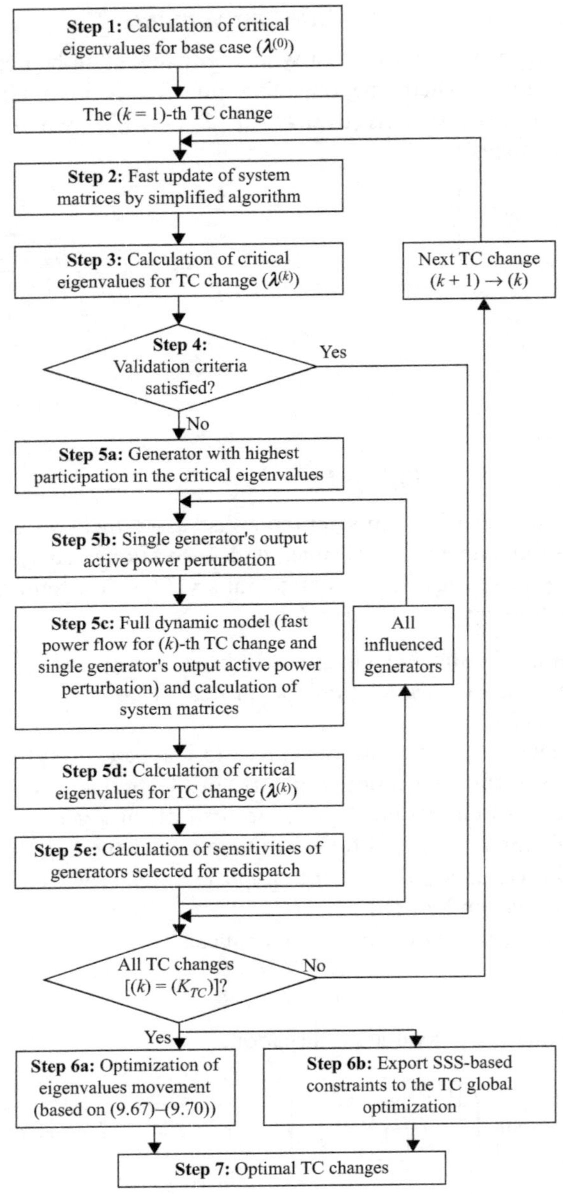

Figure 4.5 Flowchart of the scalable and tractable algorithm for fast assessment of SSS, eigenvalue sensitivities to simultaneous TC changes, and generations redispatch, as well as optimization of eigenvalues movement to satisfy SSS constraints (Fig. 1 1 2015 IEEE. Reprinted with permission from Reference 15)

If validation criteria are not satisfied, Continue with **Step 5a**, or Go to **Step 6a** or **Step 6b**.

Step 5a: Identify the generators with dominant participation (with maximum values of participation factors, given in (4.15)) in the critical oscillation modes (critical eigenvalues) ($N_G^{(k)} = U_i(N_{Gi}^{(k)}),\ i = 1,\ 2,\ \cdots,\ n_c$) [2–8].
Step 5b: Specify the generator output active power perturbations (ΔP_{Gj}^{sp}) for list of selected generators for redispatch in **Step 5a** (set of generators $N_G^{(k)}$).
Step 5c: Recalculate the full dynamic model for (k)-th TC change (fast power flow for TC change and single generator's output active power perturbation can be used [65]), and therefore calculation of system matrices.
Step 5d: Calculation of closest eigenvalues to the critical eigenvalues in base case operating condition ($\boldsymbol{\lambda}^{(0)}$) for ($k$)-th topology change and all selected generator's output active power perturbations ($\boldsymbol{\lambda}^{(k)}$).
Step 5e: Calculation of sensitivities for (k)-th TC change and generators selected for redispatch [matrix $\boldsymbol{K}_{\sigma}^{(k)}$ in (4.69)].
Step 6a: Optimization of eigenvalues movement [based on (4.67)–(4.70)].
Step 6b: Export SSS-based constraints to TC global optimization (respecting the cost minimization with satisfied transmission constraints, voltage constraints, etc.) [59, 60].
Step 7: Optimized TC changes.

Please note that the simplified algorithm (Section 4.7.1) is used for prefiltering of TC changes that have negligible effects on eigenvalues (and which we expect to be numerous in large-scale power systems) and that the role of **Step 5** is to zoom on TC changes that can have significant effects. Also, note that the TC actions can be optimized locally (**Step 6a**, using only SSS constraints), or as a part of a larger optimization (**Step 6b**). Numerical methods applied for different steps in the algorithm shown in Figure 4.5 and the computational times for these steps are discussed in Reference 15.

4.8 Illustrative test results

We developed a MATLAB script for DVS and SSS assessments, that is integrated with a professional software package for power flow, small-signal and transient analysis (DSA Tools [46]), enabling the exchange of system matrices (4.5).

The scalable and tractable algorithm for fast DVS and SSS assessment, eigenvalue sensitivities to simultaneous TC changes and generation redispatch, as well as optimization of eigenvalues movement to satisfy SSS constraints, is evaluated on two test examples:

- Small-scale (New England) 39-bus, 46-branch and 10-generator test system [9].
- Large-scale real-world (PJM) 13709-bus, 18285-branch and 2532-generator test system.

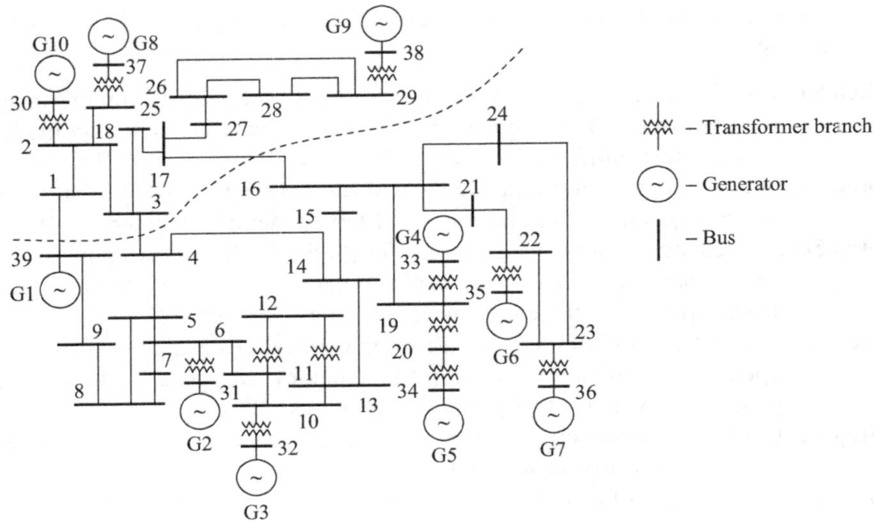

Figure 4.6 Single-line diagram for small-scale (New England) 39-bus, 46-branch and 10-generator test system

4.8.1 Small-scale (New England) 39-bus, 46-branch and 10-generator test system

Single-line diagram for analyzed small-scale (New England) test system is shown in Figure 4.6. The dynamic model of test system is described by GENROU model for all synchronous generators and ESDC1A model for all exciters [46]. This means that the dynamic model consists 110 state (10×6 for GENROU and 10×5 for ESDC1A) and 78 algebraic variables (39×2 for buses), related to dynamic elements and power flow equations, respectively.

In general, one is interested not only in the eigenvalues with positive real part but also in eigenvalues closest to the imaginary axis in the left half-plane. In SSS analysis, these eigenvalues are characterized as having a small damping ratio [ξ_i in (4.60)], where i-th ($i = 1, 2, \cdots, n$) eigenvalue is determined as $\lambda_i = \sigma_i + j\omega_i$ (e.g., $\xi_i \geq 5\%$, as shown in Figure 4.7). For the base case condition (full network topology), all eigenvalues are with negative real part ($\sigma_i \leq 0$) and with damping ratio $\xi_i \geq 5\%$. This means that the analyzed operating condition is stable for small perturbations of input variables.

We first explore the influence of changes in branch statuses to the SSS. Characteristic results are shown in Figures 4.7. On these figures the bounds for damping ratios $\xi = 7\%$ and $\xi = 5\%$ are shown in dashed and dotted lines, respectively. From presented results, we can conclude that depending on the disconnected branch, their influence on the SSS can be *negative* (Figure 4.7), *positive*, or very *small* [64]). The presented results show that the topology changes can have very different impact on modal properties in a power system: some result in a

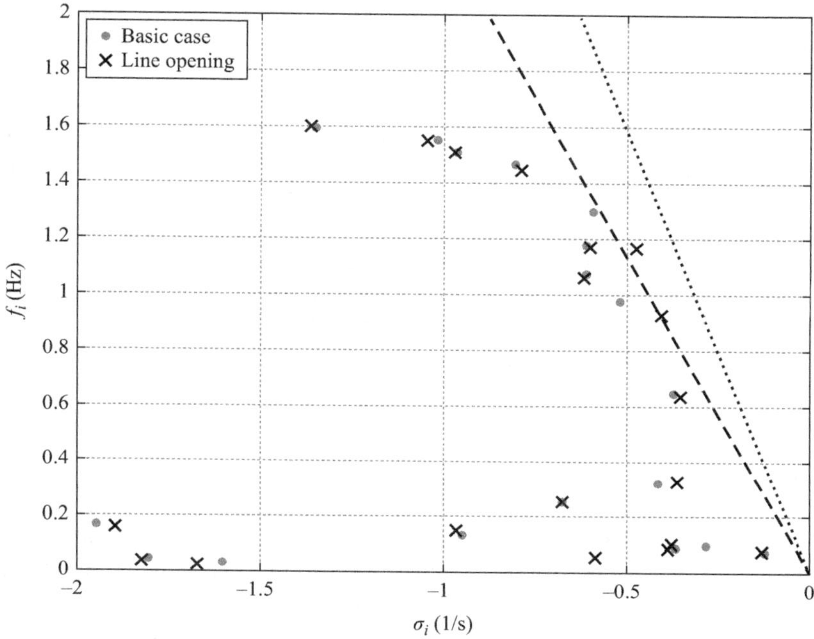

Figure 4.7 Negative influence of branch connection status (line 2–3 opening) to eigenvalue positions in complex plane

low-damped condition (Figure 4.7), while others can improve SSS [64]; similar results were presented in Reference 66.

Influence of load/generation perturbations to the eigenvalue positions is explored in Reference 64. Based on the presented results it has been concluded that these perturbations sometimes drive the power system into low-damping (e.g., $\xi < 5\%$) conditions and that the pole movement typically is almost horizontal (with constant frequency).

For small-scale (New England) test system, two characteristic cases for DVS are analyzed:

Case 1: All buses are included in source and sink areas, where total active power generation in source area is $\sum_{i \in \text{Source}} P_{Gi} = 6198.659$ MW and the total active load in sink area is $\sum_{j \in \text{Sink}} P_{Lj} = 5950.5$ MW (active power losses are $P_{\text{Losses}} = 248.159$ MW).

Case 2: Different source and sink areas (divided by dashed line in Figure 4.6), connected with transmission corridor determined by lines 1–39, 3–4 and 17–16, where the source (sink) area is on side buses 1, 3, and 17 (39, 4, and 16), where the total surplus of active power generation in source area (export to sink area by the transmission corridor), is $\sum_{i \in \text{Source}} P_{Gi} - \sum_{j \in \text{Sink}} P_{Lj} = 1620.0 - 1455.5 = 164.5$ MW.

For LIBs described by (4.26), the generator's reactive power (4.47) and bus voltage (4.48) constraints are analyzed in both cases, while in **Case 2** the constraint of active power flow by the transmission corridor (4.49) is additionally checked ($\leq$250 MW).

Tracings of five critical eigenvalues {rightmost [for λ_{max} in (4.50)] or closest-to-zero [for λ' in (4.51) and λ'' defined in Section 4.6.3], respectively} depending on the increased loading factor (ζ, with step $\Delta\zeta = 0.05$ p.u.[4]) for **Case 2** and for DAEs/AEs models are shown in Figure 4.8. (Results for **Case 1** are shown in Reference 26.) The movements of critical eigenvalues (that determines the nature of particular bifurcation) are underlined by dotted arrows in Figure 4.8. The voltage magnitudes in five most critical buses (with largest deviations) are presented in Figure 4.9, for both analyzed cases. In Figure 4.10 are shown the voltage magnitudes at ending buses in function of active power over transmission corridor (*P–V* curves). In Figures 4.8–4.10, points corresponding to different types of bifurcations are additionally marked with arrows.

The loading margins that correspond to different types of bifurcations are presented in Table 4.1. In Table 4.2 are reported the dominant state/algebraic variables and their participation factors for bifurcation points from Table 4.1.

From Figure 4.8(a) and Table 4.1, we can conclude that in both cases the HBs are critical, leading to the system's oscillatory instability. This means that loading margins for static and dynamic VS are different. In analyzed cases, the loading margin for dynamic VS is significantly lower than the static VS [0.180 p.u. vs. 0.350 p.u. (for **Case 1**) and 0.100 p.u. vs. 0.345 p.u. (for **Case 2**)]. In cases where only static VS are analyzed, the LIBs are critical and appear before SIB (for full power system Jacobian or *Q–V* sensitivities). From Table 4.2 we can conclude that the machine's flux state variables are dominant contributors for HB.

Our findings about critical HBs and loading factors (Table 4.1) suggest the presence of oscillatory instability of short-term dynamics. We verify this condition via time-domain simulations. The maximum loading operation condition in **Case 1**, Table 4.1 and identified HB (for loading margin $\zeta = 0.18$ p.u.) is subjected to a small perturbation: increase active and reactive power loads in buses 3 and 4 for $\Delta\zeta = 0.05$ p.u. ($\Delta P_{L_3} = 0.190$ p.u., $\Delta P_{L_4} = 0.295$ p.u., $\Delta P_{L_3} = 0.014$ p.u. and $\Delta P_{L_4} = 0.108$ p.u., respectively) in $t = 0.2$ s (note that these load increases are significantly smaller than the load perturbation assumed in **Case 1**, where all load buses undergo the increases). The time responses for generator active powers are shown in Figure 4.11. These results clearly indicate oscillatory instability for even a small load perturbation. This effect is clear for generators in buses 36 and 38 (these results are in agreement with the participation factors listed in Table 4.2), as well as for the slack generator (in bus 39, since the generator redispatch is not utilized).

The analyzed small-scale New England test system is well damped (for base case condition is $\xi \geq 7\%$, while for all line-opening conditions is $\xi \geq 5\%$ – see Figure 4.7). This means that the SSS constraints for line opening conditions are not

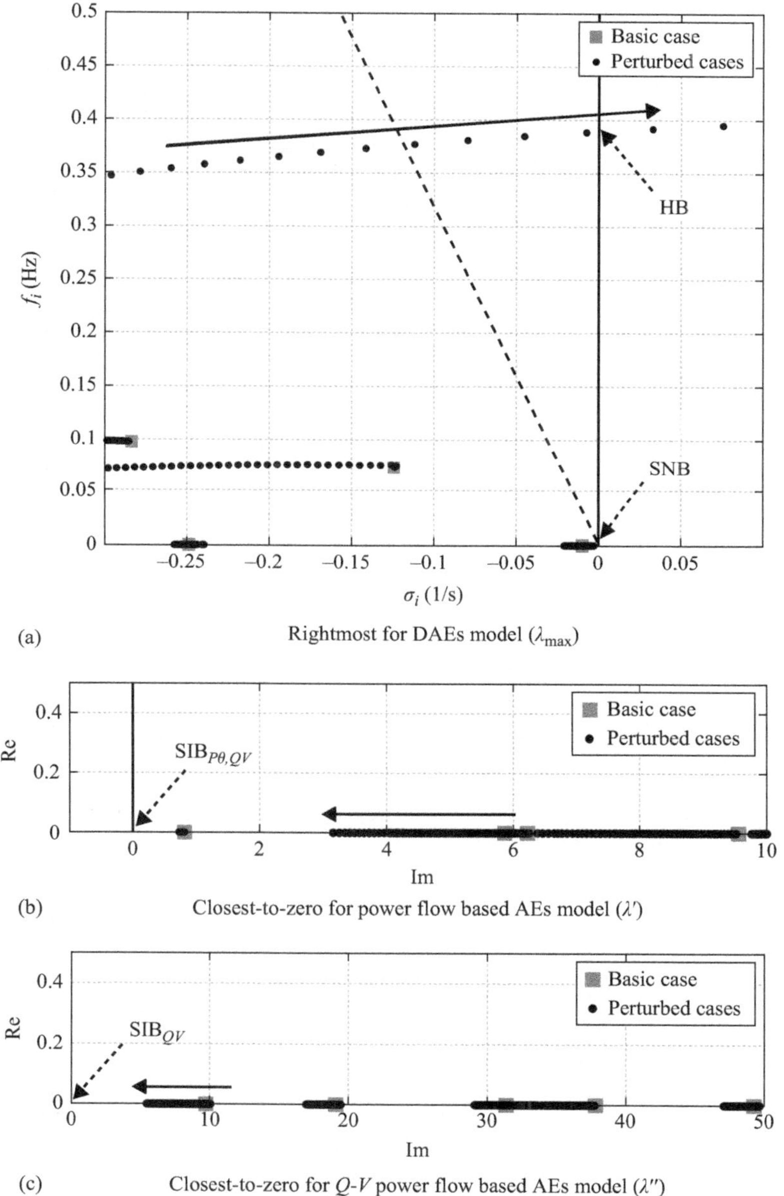

Figure 4.8 Tracings of five critical eigenvalues for ***Case 2*** *– with dashed line in panel (a) is shown damping ratio of 5% (Fig. 2 1 2015 IEEE. Reprinted with permission from Reference 15)*

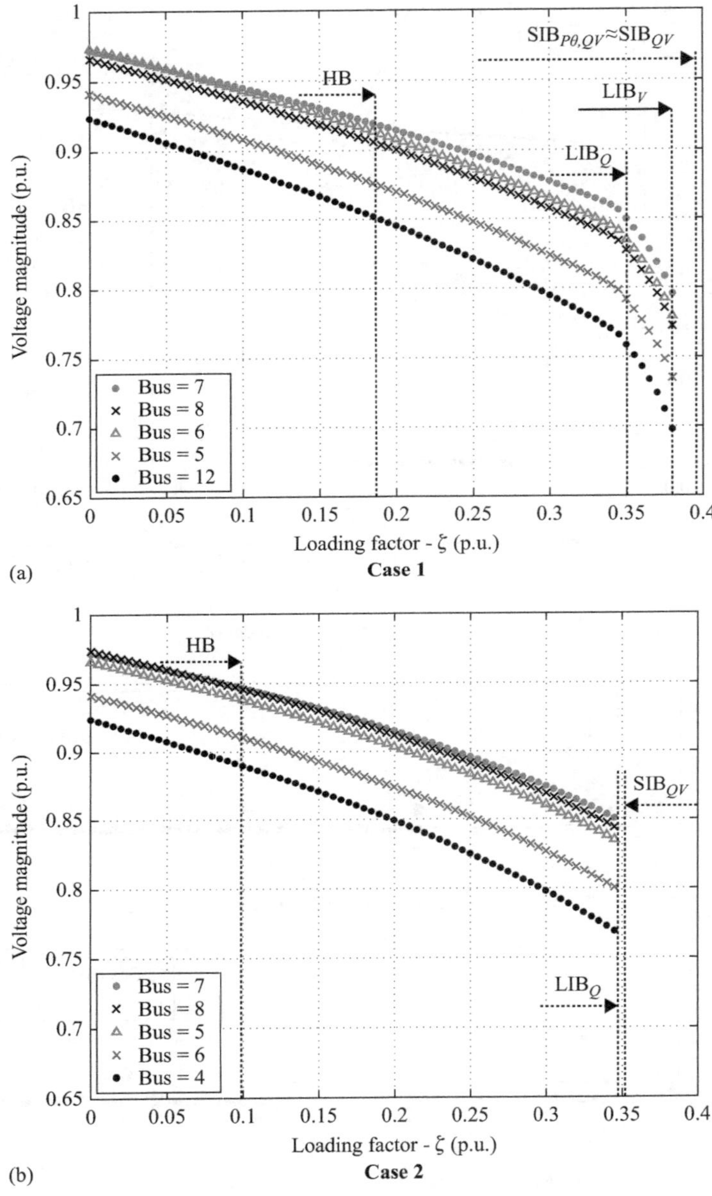

Figure 4.9 Maximum five voltage magnitude deviations in function of loading factor (P–V curve) (Fig. 3 1 2013 IEEE. Reprinted with permission from Reference 26)

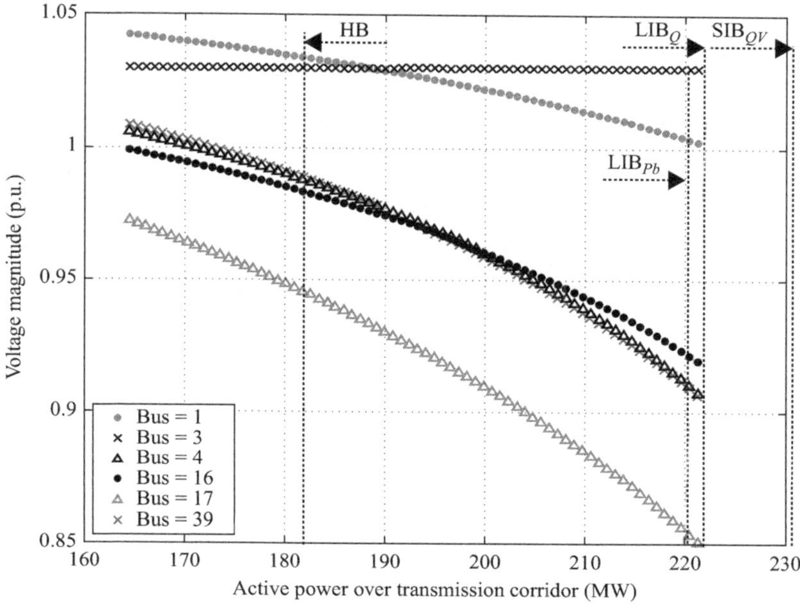

Figure 4.10 Voltage magnitudes at ending buses in function of active power over transmission corridor (P–V curves) for ***Case 2*** *from Table 4.1 (Fig. 3 1 2015 IEEE. Reprinted with permission from Reference 15)*

Table 4.1 Loading margins (in [p.u.] for different types of bifurcations)

	HB (4.27)	**SNB**	**LIB$_x$ (4.46)**	**LIB$_Q$ (4.47)**	**LIB$_V$ (4.48)**	**LIB$_{Pb}$ (4.49)**	**SIB$_{P\theta,QV}$ (4.30)**	**SIB$_{QV}$ (4.56)**
Case 1	0.180	–	–	0.350	0.380	–	≈0.395	≈0.395
Case 2	0.100	–	–	0.345	–	0.310	–	≈0.400

critical. For this reason, results of methodologies described in Section 4.7.2 and Section 4.7.3 are not shown for New England test system and will be demonstrated on large-scale real-world (PJM) test system.

4.8.2 Large-scale real-world (PJM) 13709-bus, 18285-branch and 2532-generator test system

Our aim has been to develop scalable and tractable algorithms for DVS and SSS assessments and additionally that would be able to quickly process a list of TC changes in a large power system. It appears that today there exist a serious barrier

Table 4.2 Dominantly influenced state/algebraic variables and their participation factors for bifurcation points from Table 4.1 (Table III 1 2016 IEEE. Reprinted with permission from Reference 26)

	State/Algebraic variable	Bus	Participation factor/Limit
	Case 1		
HB	GENROU, ψ_{fd}	38	1.00
	ESDC1A, Lead–lag	38	0.88
	GENROU, ψ_{fd}	36	0.23
LIB_Q	Q_G	32	800 MVAr*
LIB_V	V	7	0.7 p.u.
$SIB_{P\theta,QV}$	$P = P_G - P_L$	36	1.000
		34	0.974
		35	0.964
SIB_{QV}	$Q = Q_G - Q_L$	12	1.000
		7	0.623
		14	0.622
	Case 2		
HB	GENROU, ψ_{fd}	38	1.00
	ESDC1A, Lead-lag	38	0.90
	GENROU, ω	38	0.23
LIB_Q	Q_G	38	800 MVAr*
SIB_{QV}	$Q = Q_G - Q_L$	7	1.000
		8	0.830
		12	0.728

*Generator in bus 31 is with maximum reactive power (800 MVAr) for "basic case" and all "perturbed cases"

at approximately 10,000 state variables. While this barrier is hardware- and software-dependent, and will undoubtedly move upward in the future, our experience suggests that a customized approach is more beneficial than a frontal assault [26]:

- Full eigenvalue analysis for more than ~10,000 state variables either cannot be executed today by commercial software packages or requires unreasonable computation times.
- Critical (e.g., rightmost, poorly damped [56] and poles most sensitive to multiple input parameter changes [19]) can be calculated in a reasonable computation time (for appropriate number of requested rightmost eigenvalues) in our PJM test example using the descriptor form (4.11).
- When calculating a subset of eigenvalues (e.g., with frequency and damping ratio ranges), the user must be careful. This kind of calculation can be time consuming (for wide ranges) or some critical eigenvalues could turn out to be outside of the specified ranges.

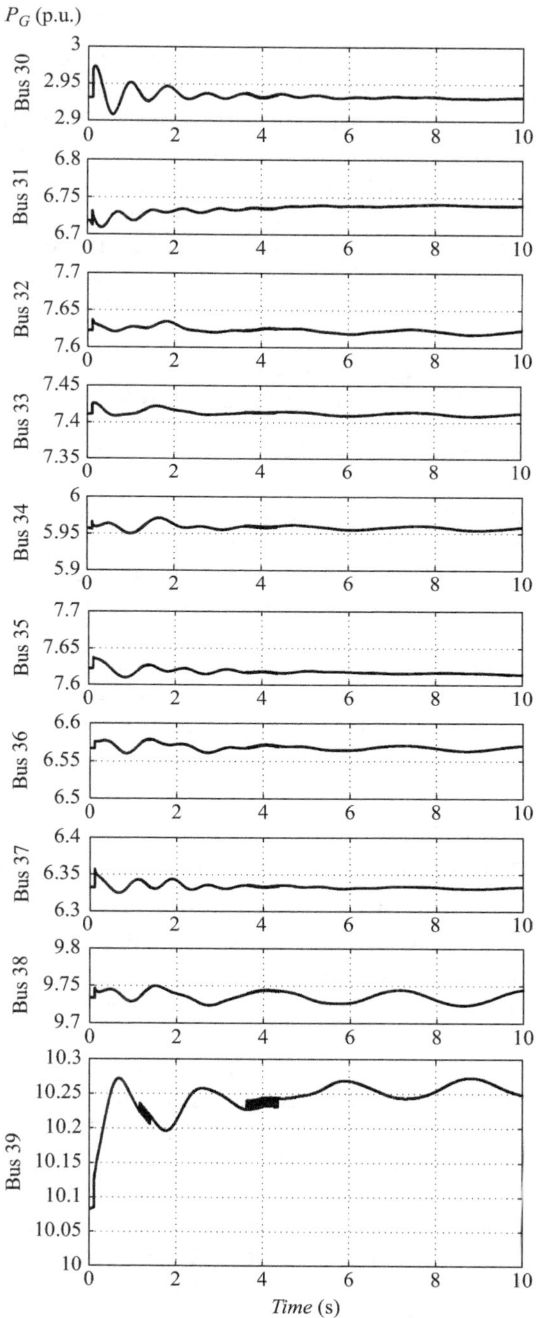

*Figure 4.11 Time response simulations for generator active powers for **Case 1** in Table 4.1 (Fig. 5 1 2016 IEEE. Reprinted with permission from Reference 26)*

Based on these considerations, our approach entails:

- For calculation of critical eigenvalues in base case operating condition (**Step 1**, flowchart in Figure 4.5), two alternatives are available: (1) a careful specification of frequency and damping ratio ranges (in commercial software packages, such as DSA Tools [46]), or (2) calculation of closest eigenvalues to the requested damping ratio (4.22), varying the frequency of small-signal oscillations (in developed MATLAB environment).
- For TC changes, calculation of perturbation of critical eigenvalues (**Step 5d**, flowchart in Figure 4.5) is performed by a very fast option "closest to the critical eigenvalues in base case operating condition" (in our MATLAB environment).
- Only critical TC changes [respecting validation criteria (4.71)–(4.73)] are recalculated with detailed dynamic model (**Step 5**, flowchart in Figure 4.5).

In this way the number of required eigenvalue calculations is largely independent of the size of the underlying system (total number of state and algebraic variables), but depends on: (1) the selected number of critical eigenvalues in base case operating condition, (2) the selected number of generators participating in critical eigenvalues, and (3) the values used for validation criteria in (4.71)–(4.73).

Main characteristics of analyzed PJM test system are 13,709 buses (13,384 in operation), 2532 generators (2113 in operation), 12,685 lines and 5600 adjustable two-winding transformers. Dynamic model of analyzed power system is composed of 12.824 state [dimension of vector $\boldsymbol{x}$ in (4.3)] and 27,648 algebraic variables [dimension of vector $\underline{V}$ in (4.4) related to dynamic elements and power flow equations].

4.8.2.1 DVS Assessment

For PJM test system, two characteristic cases for DVS were analyzed:

Case 1: All buses are included in source and sink areas, as it is relevant for all the above-mentioned bifurcation types [excluding the congested ("bottle-neck") transmission corridors, when the DVS typically determines LIB (4.49)]. The total active power generation in source area is $\sum_{i \in \text{Source}} P_{Gi} =$ 262090.5 MW and total active load in sink area is $\sum_{j \in \text{Sink}} P_{Lj} =$ 251913.51 MW (active power losses are $P_{\text{Losses}} = 10{,}176.99$ MW).

Case 2: Different source and sink areas, connected with central interface corridor, determined by 500 kV lines CONASTON–PEACHBOT (1087-605), CONEMAUG–JUNIATA (1769–833) and KEYSTONE–JUNIATA (1858–833), where sink area is on side of buses 1087, 1769 and 1858 [in areas 5 (BC) and 8 (PN) loads are increased], while the source area is on side of buses 605 and 833 [in areas 2 (PE) and 3 (PL) generations are increased]. The total surplus of active power generation in source area (export to sink area by the transmission corridor) is $\sum_{i \in \text{Source}} P_{Gi} - \sum_{j \in \text{Sink}} P_{Lj} = 9502.88 - 6855.85 = 3445.28$ MW. The constraint of transmission corridor (4.49) is $P_b \leq 6000$ MW.

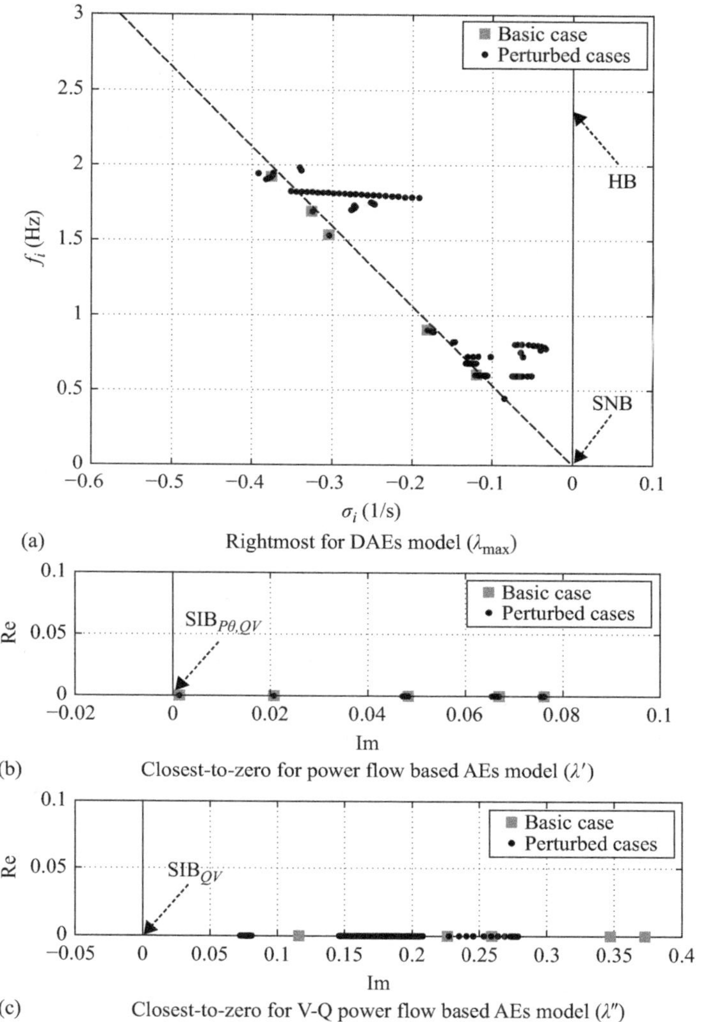

Figure 4.12 Tracings of five critical eigenvalues for ***Case 2*** *– with dashed line in panel (a) is shown damping ratio of 3% (Fig. 4 1 2015 IEEE. Reprinted with permission from Reference 15)*

Movements of five critical (rightmost) eigenvalues in function of loading factor (ζ) are shown in Figure 4.11 (for Case **2**). These results are obtained for basic network topology.

In Figure 4.12, five buses with maximal voltage magnitude deviations in function of loading factor are shown. (It is interesting to note that in **Case 1** there are buses with increased voltage magnitudes.) In Figure 4.13, the voltage magnitudes at ending buses in function of active power over transmission corridor

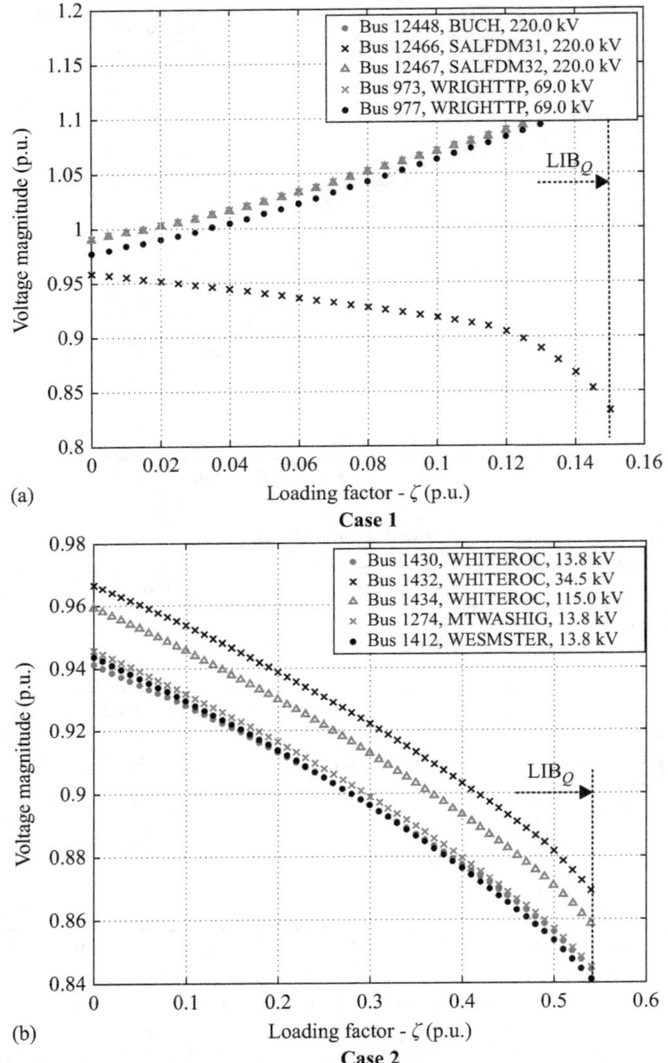

*Figure 4.13 Maximum five voltage magnitude deviations in function of loading factor (P–V curve) (**Case 1**: Fig. 7 1 2016 IEEE. Reprinted with permission from Reference 26; **Case 2**: Fig. 5 1 2015 IEEE. Reprinted with permission from Reference 15)*

(*V–P* curves) for **Case 2** are displayed. It is interesting to note that there are buses with increased voltage magnitudes. Figure 4.14 shows the number of buses that have reached voltage and generator's reactive power generation constraints as a function of the loading factor.

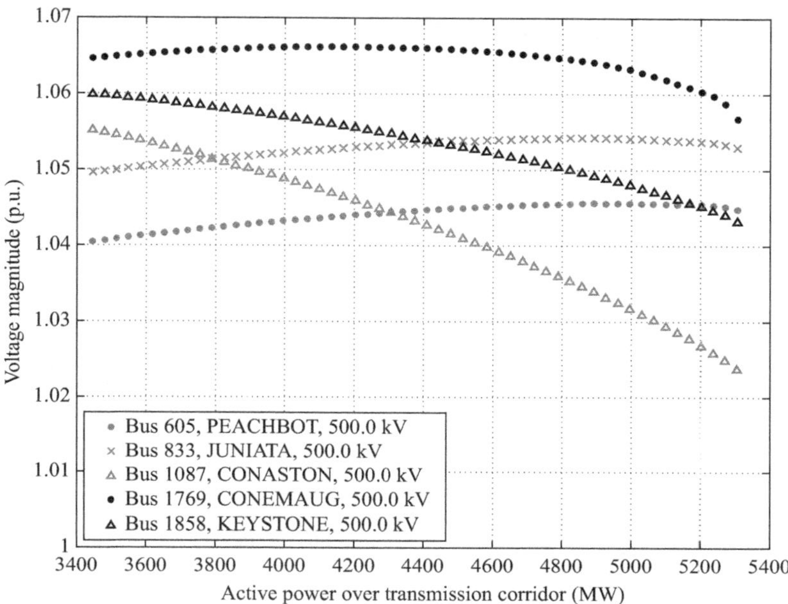

Figure 4.14 Voltage magnitudes at ending buses in function of active power over transmission corridor (P–V curves) for ***Case 2*** *(Fig. 7 1 2015 IEEE. Reprinted with permission from Reference 15)*

The obtained results can be summarized as:

- Maximum loading factor is $\zeta_{\max} = 0.150$ p.u. (**Case 1**) and $\zeta_{\max} = 0.540$ p.u. (**Case 2**, Figure 4.13).
- Damping ratio varies in the range 3.07 (for $\zeta = 0$) to 1.67 (for $\zeta = 0.510$ p.u. in **Case 1**) or to 1.92 (for $\zeta = 0.08$ p.u. in **Case 2**), implying that HB and SNB ($\lambda_{\max}$) are not critical (e.g., see Figure 4.15).
- Leftmost real eigenvalue λ' is very slowly varying with loading factor (see Figure 4.15(b) and [26]), for all analyzed loading factors, thus being so far from zero that $\text{SIB}_{P\theta,QV}$ is not critical for both analyzed cases.
- The fastest movement to the zero for λ'' is from 0.1168/0.2263 or (for $\zeta = 0$) to 0.0583/0.1416 (for $\zeta_{\max} = 0.15/0.54$ p.u.) for **Case 1** and **Case 2**, respectively, implying that this value is still so far from zero that SIB_{QV} is not critical.
- In both test cases, the LIBs are critical for DVS. Figure 4.16 shows the number of reached bus voltage and generator's reactive power constraints as function of the loading factor. From presented results we can derive the conclusion that LIB_Q are critical for DVS [over 260 (**Case 1**) and 150 generators (**Case 2**) are at their maximal reactive power generation].

Examples of TC actions in the PJM test system are described in Reference 67. The duration of these actions varies significantly: topology changes tend to last at

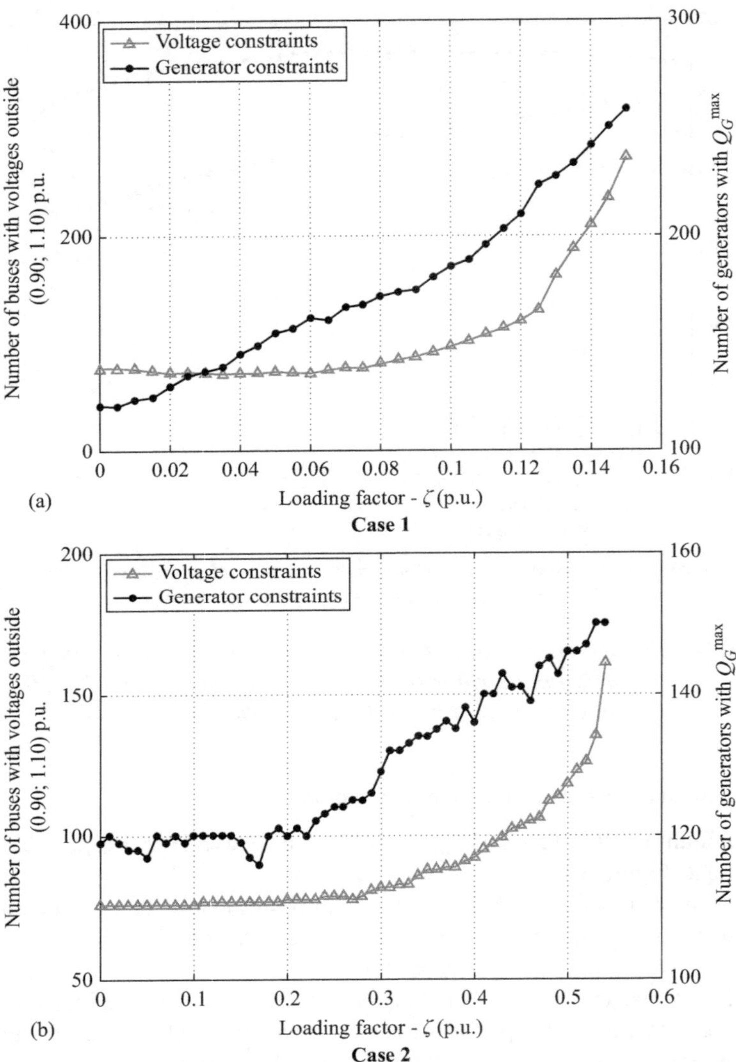

*Figure 4.15 Number of reached constraints in function of loading factor (buses with voltages outside [0.9; 1.1] p.u. and generators with maximal reactive power) (**Case 1**: Fig. 6 1 2016 IEEE. Reprinted with permission from Reference 26; **Case 2**: Fig. 7 1 2015 IEEE. Reprinted with permission from Reference 15)*

least several hours, but could also be seasonal. The rate of switching for a given piece of equipment is quite low. For example, total number of topology changes per hour in PJM has a median that varies between 0 and 3 branches opened and closed, depending on the season [59, 67].

Table 4.3 List of optimal generation redispatch actions (Table IV 1 2016 IEEE. Reprinted with permission from Reference 26)

Bus	Bus name (kV)	Unit	P_{G_0} (p.u.)	ΔP_G (p.u.)	$P_G = P_{G_0} + \Delta P_G$ (p.u.)
31	BERGEN, 18	2	67.32	−52.32	15.00
33	BERGEN, 18	1	73.13	−58.13	15.00
33	BERGEN, 18	2	74.44	−59.44	15.00
142	LINDEN, 18	1	127.28	−28.28	99.00
2072	DICKERSO, 13.8	2	84.18	−44.18	40.00
2183	CHAMBERS, 230	1	189.90	−143.90	46.00
3827	29 JOLIE, 24	1	262.71	247.29	510.00
3827	29 JOLIE, 24	2	273.40	236.60	510.00
7250	SHAMPTON, 115	1	64.09	−44.09	20.00

Table 4.4 List of optimal TC actions, maximum loadings for DVS and identified bifurcation types (Table V 1 2016 IEEE. Reprinted with permission from Reference 26)

Bus	Bus name (kV)	Bus	Bus name (kV)	Circuit	ζ_{max} (p.u.)	Bifurcation type
108	HUDSON, 230	87	ESSEX, 230	1	0.142	LIB_Q
151	MARION, 138	106	HUDSON, 138	1	0.141	LIB_Q
833	JUNIATA, 500	832	JUNIATA, 230	1	0.140	LIB_Q
1950	SHAWVILL, 230	2883	MOSHANON, 230	1	0.140	LIB_Q
2597	SMAWA, 345	2596	SMAWA, 138	1	0.141	LIB_Q
2642	BLACKOAK, 138	2813	JUNCTION, 138	1	0.141	LIB_Q
2863	MEADOWBR, 138	2862	MEADOWBR, 500	4	0.141	LIB_Q
3500	150 CALU, 345	3650	177 BURN, 345	1	0.142	LIB_Q
4634	AMOS, 765	4639	AMOS, 1.0	1	0.141	LIB_Q
4805	CAPITOLH, 138	4807	CAPITOLH, 46	1	0.141	LIB_Q
4968	DEERCREE, 138	5550	MULFBTAP, 138	1	0.142	LIB_Q
5334	KANAWHAR, 345	5335	KANAWHAR, 138	1	0.140	LIB_Q

The results of TC optimization for the selected case are shown in Tables 4.3 and 4.4.

In our simulations, the generator's ramp-up (dawn) rates are disregarded and we assume that the particularly topology change (Table 4.3) and generation redispatch actions (Table 4.4) are implemented simultaneously. In all analyzed cases, the transient instability (see algorithm in Figure 4.5) has not been identified. The load margins for DVS and critical bifurcation types are shown in last two columns of Table 4.4. As expected, the loading margin decreases with generation rescheduling and branch (line and transformer) switching; in all analyzed cases, LIBs are critical for DVS.

Numerical issues in the described algorithm are multifaceted, so it is thus clear that the size of optimization model (4.38)–(4.44) is formidable, and we cannot

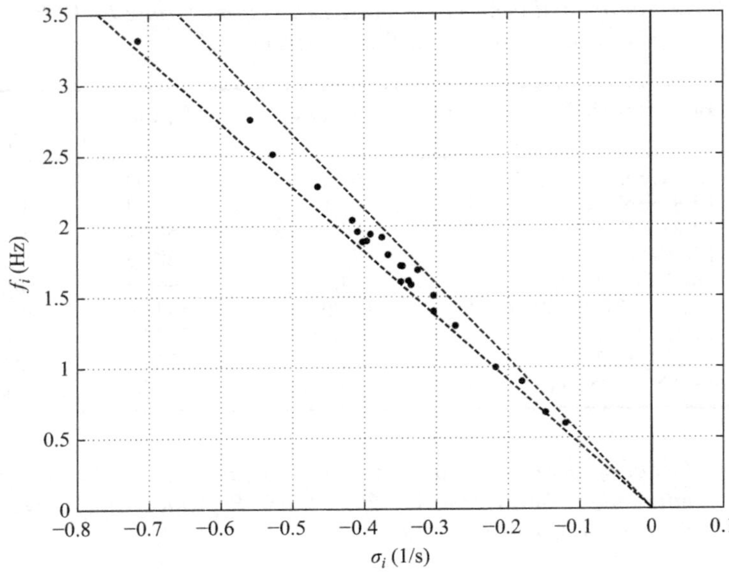

Figure 4.16 Positions of critical eigenvalues (with damping ratio between 3% and 3.5% (shown by dashed lines) for all frequencies) (Fig. 4 1 2015 IEEE. Reprinted with permission from Reference 15)

expect to solve it with reasonable computation times, given the multiple calculations in iterative optimization-based predictor–corrector algorithm for DVS tracing. Computation times for both algorithms described in Section 4.6.1 (Predictor–corrector-based algorithm) and Section 4.6.2 (Interval bisection-based algorithm) are discussed in References 15 and 26.

4.8.2.2 TC optimization with SSS constraints

The total of $n_c = 24$ critical eigenvalues (with damping ratio (ξ) between 3% and 3.5% for all frequencies) are shown in Figure 4.16. Total number of single lines or one line from multiple circuits in analyzed test system is 12225. Lines that isolate the generators from the grid (changing the size of differential part of DAEs model) are additionally removed from the list (the total of 24 lines). Therefore, the total number of line openings in analyzed list of TC changes is 12201.

All lines in list of TC changes are processed by simplified methodology for fast update of system matrices (Section 4.7.2) and calculation of perturbations for critical eigenvalues. Distribution of validation criteria (4.71)–(4.73) are shown in Table 4.5. Only 68 TC changes (0.557%) generate the low-damped eigenvalues [first validation criterion (4.80)]. Similarly, only 70 (41) TC changes [0.574% (0.336%)] have maximum eigenvalue perturbation, (4.81) [Average eigenvalues perturbation (4.82)] validation criterion larger than 0.6% (0.05%). Note that all criteria yield similar lists of critical TC changes.

Table 4.5 Distribution of validation criteria (4.71)–(4.73) for list of TC changes (TABLE IV 1 2015 IEEE. Reprinted with permission from Reference 15).

Number of low-damped eigenvalues (4.71)					
Interval of damping ratios, in (%)					
Number of TC changes					
>3.00	3.00–2.90	2.90–2.80	2.80–2.70	2.70–2.50	2.50–1.98
12133	29	21	5	7	6
Maximum eigenvalue perturbation (4.72)					
Interval of maximum eigenvalue perturbation, in (%)					
Number of TC changes					
$<10^{-5}$	10^{-5}–0.20	0.20–0.60	0.60–1.00	1.00–1.50	1.50–2.90
11231	714	186	40	23	7
Average eigenvalues perturbation (4.73)					
Interval of average eigenvalues perturbation, in (%)					
Number of TC changes					
$<10^{-5}$	10^{-5}–0.01	0.01–0.02	0.02–0.05	0.05–0.09	0.09–0.19
11487	452	115	106	38	3

In the list of 24 analyzed critical eigenvalues, only five are perturbed significantly – moving to the low-damped area (see Fig. 5 in Reference 15) for all TC changes (mainly on lower frequencies). Their details are shown in Table V in Reference 15; generator participation factors are largely unaffected by line openings. From these results, it is clear that there is the eigenvalue dominantly affected by the exciter. This eigenvalue cannot be repositioned to the well-damped region by generator output perturbations, which we had confirmed by additional simulations. This means that all line openings that make this particular eigenvalue poorly damped cannot be allowed in the TC process. Retuning of the exciter would be a preferred solution, of course, but we do not consider operation planning aspects here.

To decrease the computation time for optimization of eigenvalues movement, only these $n_c = 5$ critical eigenvalues are used (instead of the 24 shown in Figure 4.16). Using results in Table 4.5 and the first validation criterion [existence of low-damped eigenvalues, (4.71)], we selected 68 critical TC changes for calculation by the full dynamic model (**Step 5**, flowchart in Figure 4.5). Detailed results are shown in Reference 15.

The overall number of calculations of closest eigenvalues is $642 \times 5 = 3210$ [nonzero elements in matrix $\boldsymbol{K}$ (4.69)]. The size of this effort is system-dependent and mainly determined by:

1. the number of critical line openings for full dynamic model analysis (typically quite limited);
2. the number of selected critical eigenvalues for further analysis (only eigenvalues for line openings with violated validation criteria (4.71) are selected for detailed analysis); and

3. the number of generators significantly participating in the selected critical eigenvalues (our experience shows that this number is typically limited to 2−3).

Computational issues for algorithm from Figure 4.5 for fast assessment of SSS, eigenvalue sensitivities to simultaneous TC changes and generations redispatch, as well as optimization of eigenvalue movements to satisfy SSS constraints, are discussed in Reference 15. Our results verify the scalability and tractability of the algorithms described in Sections 4.7.2 and 4.7.3 for dynamic power system topology optimization, involving changes in branch status and generation redispatch, while respecting SSS constraints. The computational effort of the described approach does not increase anywhere near exponentially with problem size, rendering the approach tractable for real-world (10,000 + bus) systems. The computational cost savings of the algorithm stem from requiring a small number of efficiently calculated topology changes, from the negligible computational cost of updating sensitivities of the system and state matrices, and from the avoidance of explicit calculation of all eigenvalues for all topology changes in SSS analysis.

References

[1] Savulescu S.C. (ed.). Real-Time Stability in Power Systems: *Techniques for Early Detection of the Risk of Blackout*. 2nd edn. Switzerland: Springer International Publishing; 2014.

[2] Ajjarapu V. *Computational Techniques for Voltage Stability Assessment and Control*. New York, NY, USA: Springer-Verlag; 2006. Chapters 2–5.

[3] Cutsem T.V., Vournas C. *Voltage Stability of Electric Power Systems*. Norwell, MA, USA: Kluwer Academic Publishers; 1998. Chapters 5, 7, 9.

[4] Machowski J., Bialek J.W., Bumby J.R. *Power System Dynamics and Stability*. 2nd edn. Chichester, UK: John Wiley & Sons Ltd; 2008. Chapters 5, 8.

[5] Pal M.K. Lecture notes on power system stability. [online]. 2007. Available from https://www.google.rs/?gws_rd=cr,ssl&ei=CfU1VfqKJ4HlsgG7qoCIBQ#q=%22power+system+stability%22+Mrinal+pal [Accessed 21 Apr 2015].

[6] Kundur P. *Power System Stability and Control*. New York, NY, USA: McGraw-Hill Inc.; 1994. Chapters 12, 14, 16.

[7] Taylor C.W. *Power System Voltage Stability*. New York, NY, USA: McGraw-Hill; 1994.

[8] Ilic M., Zaborszky J. *Dynamics and Control of Large Electric Power Systems*. New York, NY, USA: Wiley; 2000.

[9] Pai M.A. *Energy Function Analysis for Power System Stability*. Norwell, MA, USA: Kluwer Academic Publishers; 1989.

[10] Vittal V. "Transient stability test systems for direct stability methods." *IEEE Transactions on Power Systems*. 1992; 7(1): 37–43.

[11] Pavella M., Ernst D., Vega R. *Transient Stability of Power Systems: A Unified Approach to Assessment and Control*. Norwell, MA, USA: Kluwer Academic Publishers; 2000. Chapter 2.

[12] Molina R.D.M, Cassano M. "Dimo's approach to steady-state stability assessment: methodology overview, numerical example, and algorithm validation" in Savulescu S.C. (ed.). *Real-Time Stability Assessment in Modern Power System Control Centers*. Hoboken, NJ, USA: John Wiley & Sons Inc; 2009. pp. 307–52.

[13] Đukić S.D., Sarić A.T. "Dynamic model reduction: an overview of available techniques with application to power systems." *Serbian Journal of Electrical Engineering (SJEE)*. 2012; 9(2): 131–169.

[14] Stanković A.M., Đukić S.D., Sarić A.T. "Approximate bisimulation-based reduction of power system dynamic models". *IEEE Transactions on Power Systems*. 2015; 30(3): 1252–1260.

[15] Sarić A.T., Stanković A.M. "Rapid small-signal stability assessment and enhancement following changes in topology". *IEEE Transactions on Power Systems*. 2015; 30(3): 1155–1163.

[16] Rogers G. *Power System Oscillations*. Norwell, MA, USA: Kluwer Academic Publishers; 2000. Chapter 3.

[17] Ilic M.D., Stanković A.M. "Voltage problems on transmission networks subject to unusual power flow patterns." *IEEE Transactions on Power Systems*. 1991; 6(2): 339–348.

[18] Smed T. "Feasible eigenvalue sensitivity for large power systems." *IEEE Transactions on Power Systems*. 1993; 8(2): 555–563.

[19] Rommes J., Martins N. "Computing large-scale system eigenvalues most sensitive to parameter changes, with applications to power system small-signal stability." *IEEE Transactions on Power Systems*. 2008; 23(2): 434–442.

[20] Eremia M., Shahidehpour M. *Handbook of Electric Power System Dynamics*. New Jersey, NJ, USA: John Wiley & Sons; 2013. Chapter 9.

[21] Messina A.R. *Inter-Area Oscillations in Power Systems*. New York, NY, USA: Springer-Verlag; 2009.

[22] Kundur P., Paserba J., Ajjarapu V., *et al.* "Definition and classification of power system stability". *IEEE-CIGRE Joint Task Force on Stability Terms and Definitions*. 2004;19(3):1387–1401.

[23] Milano M. *Power System Modelling and Scripting*. London, UK: Springer-Verlag; 2010. Chapters 5, 7.

[24] Guckenheimer J., Holmes P. Nonlinear Oscillations, *Dynamical Systems and Bifurcation of Vector Fields*. New York, NY, USA: Springer-Verlag, 1983.

[25] Perko L. *Differential Equations and Dynamical Systems*. New York, NY, USA: Springer-Verlag, 1991.

[26] Stanković A.M., Sarić A.T. "Dynamic voltage stability assessment in large power systems with topology control actions". IEEE Transactions on Power Systems. 2016; To be published.

[27] Chiang H.D., Flueck A.J., Shah K.S., Balu N. "CPFLOW: a practical tool for tracing power system steady-state stationary behavior due to load and generation variations." *IEEE Transactions on Power Systems*. 1995;10(2):623–634.

[28] Cutsem T.V., Grenier M.E., Lefebvre D. "Combined detailed and quasi steady-state time simulations for large disturbance analysis." *International Journal of Electrical Power and Energy Systems*. 2006; 28(9):634–642.

[29] Wang Q., Song H., Ajjarapu V. "Continuation based quasi-steady-state analysis." *IEEE Transactions on Power Systems*. 2006;21(1):171–179.

[30] Dobson I. "The irrelevance of electric power system dynamics for the loading margin to voltage collapse and its sensitivities." *Nonlinear Theory and Its Application*. 2011;2(3):263–280.

[31] Canizares C.A., Alvarado F.L., DeMarco C.L., Dobson I., Long W.F. "Point of collapse methods applied to ac/dc power systems." *IEEE Transactions on Power Systems*. 1992;7(2):673–683.

[32] Kim K., Schattler H., Venkatasubramanian V., Zaborsky J., Hirsch P. "Methods for calculating oscillations in large power systems". *IEEE Transactions on Power Systems*. 1997;12(4):1639–1648.

[33] Gomes S., Martins N., Portela C. "Computing small-signal stability boundaries for large-scale power systems." *IEEE Transactions on Power Systems*. 2003;18(2):747–752.

[34] Mithulananthan N., Canizares C.A. "Hopf bifurcations and critical mode damping of power systems for different static load models." *IEEE Power Engineering Society General Meeting*. Denver, CO, USA, 2004.

[35] Zhou Y., Ajjarapu V. "A fast algorithm for identification and tracing of voltage and oscillatory stability margin boundaries." *Proceedings of the IEEE*. 2005;93(5):934–946.

[36] Armenta S.M., Esquivel C.R.F., Becerril R. "A numerical study of the effect of degenerate Hopf bifurcations on the voltage stability in power systems." *Electric Power Systems Research*. 2013;101(1):102–109.

[37] Wen X., Ajjarapu V. "Application of a novel eigenvalue trajectory tracing method to identify both oscillatory stability margin and damping margin." *IEEE Transactions on Power Systems*. 2006;21(2):817–824.

[38] Ayasun S., Nwankpa C.O., Kwatny H.G. "Computation of singular and singularity induced bifurcation points of differential algebraic power system model." *IEEE Transactions on Circuits and Systems I: Regular Papers*. 2004;51(8):1525–1538.

[39] Avalos R.J., Canizares C.A., Milano F., Conejo A.J. "Equivalency of continuation and optimization methods to determine saddle-node and limit-induced bifurcations in power systems." *IEEE Transactions on Circuits and Systems I: Regular Papers*. 2009;56(1):210–223.

[40] Cutsem T.V., Vournas C.D. "Voltage stability analysis in transient and mid-term time scales." *IEEE Transactions on Power Systems*. 1996;11(1): 146–154.

[41] Greene S., Dobson I., Alvarado F.L. "Sensitivity of the loading margin to voltage collapse with respect to arbitrary parameters." *IEEE Transactions on Power Systems*. 1997;12(1):262–272.

[42] Yang D., Ajjarapu V. "A decoupled time-domain simulation method via invariant subspace partition for power system analyses." *IEEE Transactions on Power Systems*. 2006;21(1):11–18.

[43] Yorino N., Li H.Q., Harada S., Ohta A., Sasaki H. "A method of voltage stability evaluation for branch and generator outage contingencies." *IEEE Transactions on Power Systems*. 2004;19(1):252–259.

[44] Sarić A.T., Stanković A.M. "Tractable and scalable algorithm for dynamic voltage stability assessment in large-scale power systems." *IEEE PowerTECH Conference, Eindhowen, Netherlands*, 2015.

[45] Yang D., Ajjarapu V. "Critical eigenvalues tracing for power system analysis via continuation of invariant subspaces and projected Arnoldi method." *IEEE Transactions on Power Systems*. 2007;22(1):324–332.

[46] Powertech Labs Inc. (Canada). *DSA Tools – Dynamic Security Assessment Software*. Surrey, British Columbia, Canada: Powertech Labs Inc., 2013.

[47] Ajjarapu V., Lee B. "Bifurcation theory and its application to nonlinear dynamical phenomena in an electrical power system." *IEEE Transactions on Power Systems*. 1992;7(1):424–431.

[48] Canizares C.A. "On bifurcation voltage collapse and load modeling." *IEEE Transactions on Power Systems*. 1995;10(1):512–522.

[49] Canizares C.A. "Calculating optimal system parameters to maximize the distance to saddle-node bifurcations." *IEEE Transactions on Circuits and Systems I: Fundamental Theory and Applications*. 1998;45(3):225–237.

[50] Dobson I. "Observations on the geometry of saddle node bifurcation and voltage collapse in electrical power systems." *IEEE Transactions on Circuits and Systems I: Fundamental Theory and Applications*. 1992;39 (3):240–243.

[51] Canizares C.A., Mithulananthan N., Milano F., Reeve J. "Linear performance indices to predict oscillatory stability problems in power systems." *IEEE Transactions on Power Systems*. 2004;19(2):1104–1114.

[52] Dobson I., Alvarado F., DeMarco C.L. "Sensitivity of Hopf bifurcation to power system parameters." *Proceedings of the 31st IEEE Conference on Decision and Control*. 1992;3:2928–2933.

[53] Mithulananthan N., Canizares, C.A., Reeve J. "Indices to detect Hopf bifurcation in power systems". *North American Power Symposium (NAPS)*, Waterloo, USA, 2000.

[54] Zhu W., Mohler R., Spee R., Mittelstadt W., Maratukulam D. 'Hopf bifurcations in a SMIB power system with SSR'. *IEEE Transactions on Power Systems*. 1996; 11(3): 1579–84.

[55] Cheng L. 'A comprehensive invariant subspace-based framework for power system small-signal stability analysis'. *Graduate Theses and Dissertations*. Paper 10109. [online]. 2011. Available from http://lib.dr.iastate.edu/etd/10109 [Accessed 20 Apr 2015].

[56] Rommes J., Martins N., Freitas F.D. 'Computing rightmost eigenvalues for small-signal stability assessment of large-scale power systems'. *IEEE Transactions on Power Systems*. 2010; 25(2): 929–38.

[57] Chakraborty K., Chakrabarti A. *Soft Computing Techniques in Voltage Security Analysis*. New Delhi, India: Springer, 2015. chapters 3 and 4.

[58] Wolsey L.A., Nemhauser G.L. *Integer and Combinatorial Optimization*. New York, NY, USA: John Wiley & Sons Inc, 1999. chapter I.1.

[59] Goldis E.A., Li X., Caramanis M.C., Rudkevich A., Ruiz P.A. 'AC-based topology control algorithms (TCA) – A PJM historical data case study'. *Presented at 48th Hawaii International Conference of System Sciences*, Kauai, HA, USA, 2015.

[60] Ruiz P.A., Foster J.M., Rudkevich A., Caramanis M.C. 'Tractable transmission topology control using sensitivity analysis'. *IEEE Transactions on Power Systems*. 2012; 27(3): 1550–59.

[61] O'Neill R.P., Baldick R., Helman U., Rothkopf M.H., Stewart W. 'Dispatchable transmission in RTO markets'. *IEEE Transactions on Power Systems*. 2005; 20(1): 171–79.

[62] Perunicic B., Ilic M., Stanković A.M. 'Short-time stabilization of power systems via line switching'. *Proceeding of the IEEE International Symposium on Circuits and Systems*. 1988. 1: 917–21.

[63] Wood A.J., Wollenberg B.F. *Power Generation, Operation and Control. 2nd edn. New York, NY, USA*: John Wiley & Sons Inc, 1996. *chapter 11.*

[64] Stanković A.M., Sarić A.T. 'Fast assessment of eigenvalue sensitivities to topology changes and injection perturbations'. *Presented at the IEEE PowerTECH Conference, Session 'Loads and Flows Modeling'*. Grenoble, France, 2013.

[65] Ruiz P.A., Sauer P.W. 'Voltage and reactive power estimation for contingency analysis using sensitivities'. *IEEE Transactions on Power Systems. 2007*; 22(2): 639–47.

[66] Chen Y., Fuller J., Diao R., Zhou N., Huang Z., Tuffner F. 'The influence of topology changes on inter area oscillation modes and mode shapes'. *Presented at the 2011 IEEE Power and Energy Society General Meeting*. Detroit, MI, USA, 2011.

[67] 'Switching Solutions'. [online]. 2015. Available from https://www.pjm.com/markets-and-operations/etools/oasis/system-information/switching-solutions.aspx [Accessed 20 Apr 2015].

Chapter 5

A fuzzy-based data mining paradigm for on-line optimal power flow analysis

Vincenzo Loia[1], *Stefania Tomasiello*[2] *and Alfredo Vaccaro*[3]

5.1 Introduction

Smart grids are considered as one of the most effective answers to the need of reliable, economic, and sustainable electricity services. The smart grids are conceived as a fusion of the energy, information, and communication infrastructures, which is obtained by designing integrated management and protection tools, able to handle heterogeneous and complex problems ranging from network optimization to security issues. In particular, issues such as grid efficiency improvement, flexible load supply, demand side management, emission control, and optimal network regulation can be addressed by a smart management system, which aims at acquiring and processing the available set of information describing the actual smart grid operation state. As easily understandable, this computing process is a very complex and time-intensive task, since it requires the periodic estimation of the power system state, the analysis of the massive data streams generated by the grid sensors and the repetitive solution of large-scale optimization problems, which are complex, nonlinear, and NP-hard problems. Moreover, in order to provide the grid operators with updated information to better understand and reduce the impact of system uncertainties associated with load and generation variations (e.g., in solar and wind power sources), the required computation times should be fast enough [1].

This goal can be achieved by means of computing paradigms which can support rapid power systems analysis in the typical context of "big data" [2–4].

It is worthwhile to point out that this may also apply to optimal power flow (OPF) analysis, which is widely used for solving many complex power system operation problems (e.g., network reconfiguration, optimal power dispatch, voltage control). In fact, one of the challenging topics of the current literature about OPF

[1]Department of Management and Innovation Systems, University of Salerno, Fisciano, Italy
[2]Department of Information Engineering, Electrical Engineering and Applied Mathematics, University of Salerno, Fisciano, Italy
[3]Department of Engineering, University of Sannio, Benevento, Italy

analyses is defining effective paradigms for reducing the complexities of the solution algorithms, which are usually computational expensive being based on iterative numerical techniques.

In this sense, alternative formalization of the problem equations, soft-computing-based solution techniques, and distributed processing architectures have been proposed in the literature [5, 6]. Actually, some approaches try to reduce the complexity and improve the convergence of OPF solution algorithms by defining more effective iterative schemes (e.g., modified Newton–Raphson and trust-region interior-point methods [7]) and/or applying advanced decomposition techniques (e.g., factorized load flow, Lagrangian relaxation, augmented Lagrangian decomposition [8, 9]). Other methods in the current literature are based on computational intelligence techniques such as genetic algorithms [10, 11], fuzzy logic programming [12, 13] and neural networks [14, 15]. Domain decomposition techniques have been recently proposed in References 16 and 17 in order to parallelize the solution algorithm.

It is also true that some techniques are able to find at least locally optimal solutions to large-scale OPF problems with a reasonable computational burden (e.g., [18, 19]).

However, even if such techniques often find globally optimal solutions [18], they may fail to converge or converge just to a local optimum. This is the reason why approaches based on the relaxation of the OPF problem have been proposed (e.g., see [20]), but they have not been checked for large-scale OPF problems.

In Reference 21, in order to avoid OPF calculations, a regression-based control scheme is proposed to determine the optimal settings of thyristor-controlled series compensators in the presence of renewable energy resources. However, even this approach has not been used for large-scale OPF problems.

The present chapter proposes a fuzzy transform (F-transform)-based method, aimed at reducing the cardinality of OPF problems in smart grids, by obtaining approximate solutions with a certain accuracy.

F-transform is a fuzzy approximation technique proposed by Perfilieva [22]. It states a functional dependency as a linear combination of basic functions. There are many applications of the F-transform in image processing, e.g., [23]–[27]. There are also applications in data analysis [28, 29] and in time series analysis [30–32].

In this chapter, the F-transform is first used for reducing the cardinality of a knowledge-base, which includes the relevant matrices of the historical power system states and the corresponding OPF solutions. More precisely, we present a two-stage computational paradigm, namely, for offline operations and online operations. In the offline state, the proposed approach classifies the F-transform of the knowledge-base in a proper number of clusters and identifies, for each cluster, a local regression model describing the multidimensional nonlinear mapping between the input and the output vectors. In the online state, it determines the cluster to which the vector of the measured smart grid state belongs and applies the corresponding local regression model to approximate the OPF solution. Finally, the effectiveness of this approximate solution is assessed by checking the problem constraints satisfaction and, if the constraints are not satisfied, the OPF problem is rigorously solved by using the approximate solution as an initial estimation. The obtained solution, jointly with the corresponding given vector of measurements, is then used to adjourn the knowledge-base.

This computing paradigm is expected to be fast and reliable. Quickness is assured by the adoption of F-transform, which allows to get an approximate OPF solution through a knowledge-base with a significantly reduced size, avoiding iterative and time-consuming algorithms. Reliability derives from the online checking of the approximate OPF solution, which invokes a rigorous solution algorithm only if a feasible OPF solution has not been determined.

The effectiveness of the proposed framework is demonstrated by means of numerical results obtained for realistic power networks, namely, the IEEE 118-bus test system and the 2383-bus Polish system, for various operating conditions.

The chapter is sectioned as follows. Section 5.2 presents a brief review of the OPF problem. In Section 5.3, the theoretical foundations and the main features of the proposed framework are presented. In Section 5.4, detailed numerical results are discussed. Finally, Section 5.5 summarizes the main conclusions and contributions of the chapter.

5.2 Problem formulation

5.2.1 Optimal power flow analysis

An OPF problem can be formulated as a nonlinear and nonconvex optimization problem. It aims at identifying the value of decision variables, including the control and the state variables, that minimizes one or more objective functions, by satisfying both equality and inequality constraints.

Consider an n_b-bus power system. Let $\mathbf{u}$ be the vector of the control variables, $\mathbf{x}$ the vector of state variables, $\mathbf{f}(.)$ a q-dimensional objective function vector, $\mathbf{g}(.)$ and $\mathbf{h}(.)$ the p-dimensional and r-dimensional vectors representing the problem constraints, respectively. Then the problem can be formalized in a compact form as follows:

$$\begin{aligned} \min_{(\mathbf{x},\mathbf{u})} \quad & \mathbf{f}(\mathbf{x},\mathbf{u}) \\ \text{s.t.} \quad & \mathbf{g}(\mathbf{x},\mathbf{u}) = 0 \\ & \mathbf{h}(\mathbf{x},\mathbf{u}) < 0 \end{aligned} \tag{5.1}$$

The objective functions $\mathbf{f}(.)$ in (5.1) can take several forms according to the specific application domain. For example, it can be referred to the minimization of the production costs, the minimization of the transmission line losses, the minimization of the voltage deviations and so on.

The vector of state variables $\mathbf{x}$ in it may be represented by some quantities, such as the voltage magnitude at load buses and the reactive power output of the generators. Similarly, the vector of control variable $\mathbf{u}$ may be mainly written in terms of some quantities, such as the active power output of the generators and the voltage magnitude at the generator buses.

Equality constraints express the nonlinear power flow equations, considering as state variables the voltage magnitude and phase angle at load buses, the voltage phase angle and the reactive power generated at the generation buses, and the active and reactive power generated at the slack bus. Instead, the inequality constraints describe the network operating constraints, which include the maximum allowable power flows for the power lines, the minimum and maximum allowable limits for most control variables, such as generator voltages, and for some dependent variables, such as bus voltage limits.

Many important power systems operation problems can be formalized by an OPF problem. In this context, the optimal active power dispatch analysis is recognized as one of the most fundamental tool, since it aims at assessing the optimal output of a number of power generators which meets the system load, at the lowest possible cost, and assures a secure a reliable power system operation.

Let $N_\mathrm{b} = \{1, \ldots, n_\mathrm{b}\}$ be the set of all buses and N_G the set of generator buses. The overall problem can be formalized by the following constrained nonlinear optimization programming problem [5]:

$$\begin{aligned}
&\min_{(V_i,\delta_i\,\forall i\in[1,N],P_{G_j},Q_{G_j}\forall j\in[1,N_G])} \sum_{i=1}^{N_G}\left(\alpha_i + \beta_i P_{G_i} + \gamma_i P_{G_i}^2\right)\\
&\text{s.t.}\;\; P_{G_i} - P_{D_i} - V_i\sum_{j=1}^{N} V_j Y_{ij}\cos(\omega_{ij}) = 0, \quad \forall i \in N_b\\
&Q_{G_j} - Q_{D_j} - V_j\sum_{k=1}^{N} V_k Y_{jk}\sin(\omega_{jk}) = 0, \quad \forall j \in N_b\\
&V_{i,\min} \le V_i \le V_{i,\max}, \quad \forall i \in N_b\\
&Q_{Gi,\min} \le Q_{Gi} \le Q_{Gi,\max}, \quad \forall i \in N_G\\
&P_{Gi,\min} \le P_{Gi} \le P_{Gi,\max}, \quad \forall i \in N_G
\end{aligned} \tag{5.2}$$

by having assumed $\omega_{ij} = \delta_i - \delta_j - \theta_{ij}$ and where:

- δ_i is the voltage angle at node i;
- θ_{ij} is the phase angle of the ij^{th} element of the bus admittance matrix;
- α_i, β_i, and γ_i are the cost coefficients of the ith generator;
- P_{G_i} and Q_{G_j} are the real and reactive power generated at i^{th} and j^{th} bus, respectively;
- P_{D_i} and Q_{D_j} are the real and reactive power demanded at i^{th} and j^{th} bus, respectively;
- V_i is the ith bus voltage;
- Y_{ij} is the ij^{th} element of the bus admittance matrix;

- $V_{i,\min}$ and $V_{i,\max}$ are the minimum and maximum allowable limits for the voltage magnitude at i^{th} bus, respectively;
- $P_{Gi,\min}$ and $P_{Gi,\max}$ are the minimum and maximum active power limits for the ith generator, respectively;
- $Q_{Gi,\min}$ and $Q_{Gi,\max}$ are the minimum and maximum reactive power limits for the ith generator, respectively.

This mathematical formalization can easily be extended to solve the power flow analysis, which is another relevant problem in modern power system operation.

Let N_{PV} be the set of voltage-controlled buses, N_P the set of buses where the injected active power is fixed, and N_Q the set of buses where the injected reactive power is fixed.

The power flow problem can be stated as a particular instance of the OPF problem, as follows [33]:

$$\min_{(V_i,\delta_i,V_{aj},V_{bj},Q_{G_j})} \sum_{i\in N_P} \left(P_i^{SP} - P_i\right)^2 + \sum_{j\in N_Q} \left(\hat{Q}_j^{SP} - \hat{Q}_j\right)^2$$

$$\text{s.t.} \quad P_i = V_i \sum_{j=1}^{N} V_j Y_{ij} \cos(\omega_{ij}), \quad \forall i \in N_P$$

$$Q_j = \hat{V}_j \sum_{k=1}^{N} V_k Y_{jk} \cos(\omega_{jk}), \quad \forall j \in N_Q \tag{5.3}$$

$$\hat{V}_i = V_{i_0} + V_{ai} - V_{bi} \quad \forall i \in N_{PV}$$

$$0 \leq (Q_i - Q_{i,\min}) \perp V_{ai} \geq 0, \quad \forall i \in N_{PV}$$

$$0 \leq (Q_{i,\max} - Q_i) \perp V_{bi} \geq 0, \quad \forall i \in N_{PV}$$

$$V_i, V_{ai}, V_{bi} \geq 0, \quad \forall i \in N_{PV}.$$

Although many classes of programming algorithms, such as nonlinear programming [34], quadratic programming [35, 36], and linear programming [37], have been proposed to solve these problems, their application in smart grids optimization may reveal some limitations, which mainly derive from their limited capability to solve large-scale problems, and the reduced capability in computing optimal solutions in near real time.

On the other hand, the availability of massive datasets reinforced by the high pervasion of communication and information technologies in modern power systems is pushing the research toward the perspective of suitable techniques for big data, here included data analytic techniques [38, 39], distributed inferential paradigms [40], and cooperative computing [41]. These novel algorithms are potentially suitable to deal with complex optimization problems in the big data setting, overcoming the intrinsic limitations of traditional solution methods, which have not been conceptualized to large-scale optimization problems. A comprehensive review on the parallel and distributed optimization algorithms based on alternating direction method of multipliers for solving big data optimization problem in a smart grid domain is presented in Reference 41. The latter describes the computational schemes, the convergence properties, and the software deployment issues, providing detailed numerical results for several case studies, which include robust power system state estimation, distributed smart grid management, and security-constrained OPF. In the smart grid domain, the large volume of the data sets makes data collection, storage and processing a very complex and demanding task. Hence, effective tools aimed at reducing the size and the cardinality of smart grids data may be very beneficial.

It should be also pointed out that smart grids data sets are often corrupted by noisy, incomplete, and heterogeneous data, which can affect the performance of inferential optimization algorithms, compromising their convergence. Thus, the design of advanced signal-processing techniques for noise rejection, and anomaly detection represents an essential prerequisite for the application of knowledge-based optimization frameworks.

Finally, the solution of OPF problems in a smart grid domain often requires the compliance with strictly time constraints. Hence an approximated solution, if quickly computed, is often more useful than a high quality one, which requires more computation times.

5.3 The proposed approach

To address the need for robust and fast OPF solutions in the context of modern smart grids, we propose a computational paradigm based on the approximation properties of the F-transform. The underlying principle is that, in practical applications, optimization algorithms are often invoked to solve OPF problems in power system configurations which are not too far different from previously encountered ones. Thus, the idea is to extract actionable intelligence from historical OPF solutions in order to solve OPF problems, avoiding unnecessary calculations for similar smart grid states.

Armed with such a vision, the proposed approach employs a database where the most relevant historical solutions of the OPF problem in the form of input/output data samples (i.e., power system states and the corresponding OPF solutions) are stored. These data can be intended as the knowledge-base of a system, which tries to infer the nonlinear, multidimensional mapping between the power system state and the OPF solution. In order to reduce the complexity of this identification problem, which could become intractable, since the size of the input and output vectors could be very large (i.e., several thousands for realistic power distribution systems) and the number of sample vectors increasing in time, a computational algorithm based on F-transform is designed. This algorithm allows to reduce the cardinality of the knowledge-base, leading to a sensible reduction of both the storage and the processing time to achieve an approximate OPF solution. This is obtained by first classifying the F-transformed knowledge-base in a proper number of clusters and identifying, for each cluster, a regression model describing the local mapping between the corresponding input and the output vectors. Then, during online operations, the predictions are computed by means of the most suitable model for a given vector of a measured state, by considering the distance between this vector and the cluster centers, and the feasibility of the final approximate solution is assessed by checking the problem constraints satisfaction. If a feasible solution cannot be determined, then the predicted output is refined by an iterative OPF solution algorithm and the corresponding computed solution, jointly with the vector of measurements, is then used for updating the knowledge-base. The main features of the proposed methodology are summarized in the following subsections.

5.3.1 F-transform: an overview

Before introducing the F-transform, the notion of fuzzy partition has to be recalled. Let $I = [a, b]$ be a closed interval and $x_1, x_2, \ldots, x_n$, with $n \geq 3$, be points of I, called nodes, such that $a = x_1 < x_2 < \ldots < x_n = b$. A fuzzy partition of I is defined as a sequence $A_1, A_2, \ldots, A_n$ of fuzzy sets $A_i : I \to [0, 1]$, with $i = 1, \ldots, n$ such that

- $A_i(x) = 0$ if $x \notin (x_{i-1}, x_{i+1})$;
- A_i is continuous and has its unique maximum at x_i, where $A_i(x_i) = 1$;
- $\sum_{i=1}^{n} A_i(x) = 1, \quad \forall x \in I.$

The fuzzy sets $A_1, A_2, \ldots, A_n$ are called basic functions. They form an uniform fuzzy partition, if the nodes are equidistant, and they can be triangular shaped or not. The norm of a uniform fuzzy partition is $h = (b - a)/(n - 1)$.

Usually the sinusoidal-shaped basic functions are used:

$$A_j(x) = \begin{cases} \dfrac{1}{2}\cos\left(\dfrac{\pi(x - x_j)}{(x_{j+1} - x_j)} + 1\right), & x \in [x_j, x_{j+1}] \\ \dfrac{1}{2}\cos\left(\dfrac{\pi(x - x_j)}{(x_j - x_{j-1})} + 1\right), & x \in [x_{j-1}, x_j] \\ 0, & \text{otherwise.} \end{cases} \tag{5.4}$$

Let $\mathbf{D}$ be an $N \times M$ data matrix, n and m two integers so that $n < N$ and $m < M$, $\{A_1, \ldots, A_n\}$ and $\{B_1, \ldots, B_m\}$ two fuzzy partitions.

The discrete F-transform of $\mathbf{D}$ with respect to $\{A_1, \ldots, A_n\}$ and $\{B_1, \ldots, B_m\}$ is an $n \times m$ matrix $\mathbf{F}$, whose elements are:

$$F_{kl} = \frac{R_{kl}}{S_{kl}} \qquad k = 1\ldots, n \quad l = 1, \ldots, m \tag{5.5}$$

being:

$$\mathbf{R} = \mathbf{A}^T\mathbf{D}\mathbf{B} \tag{5.6}$$

$$\mathbf{S} = \mathbf{A}^T\mathbf{J}\mathbf{B} \tag{5.7}$$

where $\mathbf{A}$ and $\mathbf{B}$ are the matrices with elements $A_k(i)$ and $B_l(j)$, respectively, and $\mathbf{J}$ is an $N \times M$ matrix with all unit elements.

Figure 5.1 summarizes how the discrete F-transform works.

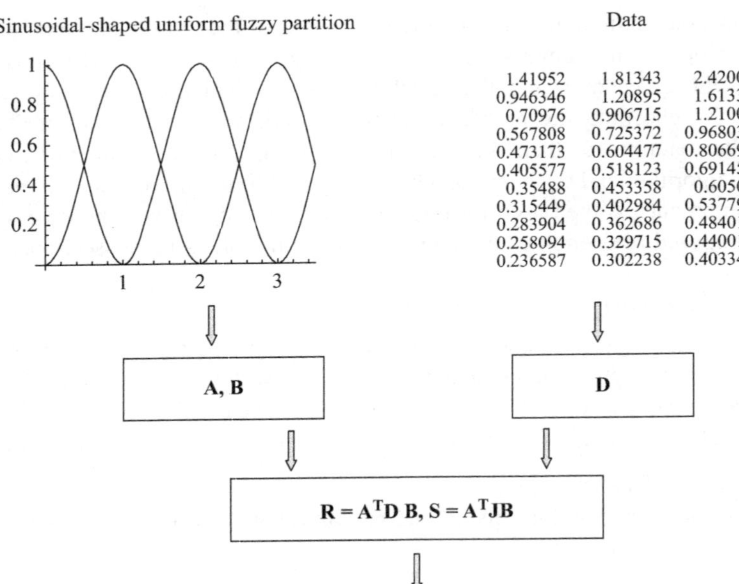

Figure 5.1 The discrete F-transform

In particular, if $\{A_1, \ldots, A_n\}$ and $\{B_1, \ldots, B_m\}$ are two uniform fuzzy partitions, with norm h_A and h_B, respectively, the elements of the matrix **S** become for $i = 2, \ldots, n-1$ and $j = 2, \ldots, m-1$ [42]:

$$S_{ij} = h_A h_B, \tag{5.8}$$

$$S_{i1} = S_{im} = S_{1j} = S_{nj} = \frac{h_A h_B}{2}, \tag{5.9}$$

$$S_{11} = S_{1m} = S_{n1} = S_{nm} = \frac{h_A h_B}{4}. \tag{5.10}$$

From matrix **F**, one can obtain an approximate $\mathbf{D}^F$ matrix by means of the discrete inverse F-transform, which is given by:

$$\mathbf{D}^F = \mathbf{AFB}^T \tag{5.11}$$

It is worthwhile to observe that each element D_{ij} of the matrix **D** may be intended as the value of a function $f : [a, b] \times [c, d] \to \mathbb{R}$ at a point (x_i, y_j).

By keeping this in mind and assuming a uniform fuzzy partition, it is possible to prove that for each $i = 1, \ldots, n$ and $j = 1, \ldots, m$ one has [42]:

$$F_{ij} = f(x_i, y_j) + O(\max(h_A^2, h_B^2)) \tag{5.12}$$

being h_A and h_B the norm of the partition in the x and y directions, respectively.

According to (5.12), the direct F-transform can approximate any function with an error depending on the square of the maximum norm of the partitions in the x and y directions.

In force of this result, in what follows only the direct F-transform will be used.

5.3.2 The offline stage

The offline stage (Figure 5.2a) is aimed at finding the relations, i.e., the approximating functions, between some measurements and the optimal settings of the problem defined in Section 2.

Let $\mathbf{X}$ be an $N \times M$ matrix, of which rows are the power system state vectors $\mathbf{X}_i$, here included the active and reactive powers measured at each network bus, and $\mathbf{Y}$ an $N \times P$ matrix, of which rows are the vectors of the corresponding OPF solutions $\mathbf{Y}_i$. Let $\mathbf{C}$ be an $N_c \times M$ matrix, of which rows are the vectors $\mathbf{C}_k$ representing substantially the centers of N_c clusters, grouping similar state vectors $\mathbf{X}_i$. Finally, let $\mathbf{Y}^C$ be the $N_c \times P$ matrix, of which rows are the vectors $\mathbf{Y}_k^C$ of the rigorous OPF solution related to the input $\mathbf{C}_k$.

For the remainder of this work, the matrices $\mathbf{C}$ and $\mathbf{Y}^C$ will represent the query matrix and the validation matrix, respectively.

Notice that $P << M$ and $N_c << N$.

Our (offline) computational scheme can be summarized by the following algorithm:

1. Compute the discrete F-transform (see Eq. (5.5)) of the matrices $\mathbf{X}$, $\mathbf{Y}$, $\mathbf{C}$, i.e., in the order of the $n \times m$ matrix $\mathbf{F}^X$, the $n \times P$ matrix $\mathbf{F}^Y$ and the $N_c \times m$ matrix $\mathbf{F}^C$, with $n < N$ and $m < M$;
2. For each value $k = 1, \ldots, N_c$

   2.1) compute the Euclidean norm

$$d(\mathbf{F}_i^X, \mathbf{F}_k^C) = \|\mathbf{F}_i^X - \mathbf{F}_k^C\|, \tag{5.13}$$

   for $i = 1, \ldots, n$;

   2.2) construct the set of r vectors, with $r \leq n/N_c$

$$Y_F^{(k)} = \{\mathbf{F}_i^Y : 1 \leq i \leq n, d(\mathbf{F}_i^X, \mathbf{F}_k^C) \leq \varepsilon\}, \tag{5.14}$$

   being ε a fixed though arbitrary small real number;

3. for $\mathbf{F}_j^{Y,(k)} \in Y_F^{(k)}, j = 1, \ldots, r$, find the mapping

$$\hat{\mathbf{Y}}_k = \beta(\mathbf{F}_1^{Y,(k)}, \ldots, \mathbf{F}_r^{Y,(k)}), \qquad k = 1, \ldots, N_c. \tag{5.15}$$

being β an unknown functional form to be determined by a regression analysis, i.e., by minimizing the deviations between $\hat{\mathbf{Y}}_k$ and the vector $\mathbf{Y}_k^C$ of the rigorous OPF solution related to the input $\mathbf{C}_k$.

Notice that the set $Y_F^{(k)}$ is substantially the cluster with center $\mathbf{F}_k^C$.

If a linear regression model is adopted, then for any $k = 1,\ldots,N_c$

$$\hat{\mathbf{Y}}_k = \sum_{j=1}^{r} a_{kj}\mathbf{F}_j^{Y,(k)}, \qquad k = 1,\ldots,N_c. \tag{5.16}$$

i.e., in compact form the regression problem can be formulated as follows:

$$\min_{\mathbf{a}} \| \mathbf{a}_k \mathbf{F}^{Y,(k)} - \mathbf{Y}_k^C \|^2 \tag{5.17}$$

where $\mathbf{a}_k$ is the vector of the unknown real parameters a_{kj}.

Let $\overline{\mathbf{Y}}^{(k)}$ denote the mean of the r vectors $\mathbf{Y}_j{}^{(k)}$, i.e., $\overline{\mathbf{Y}}^{(k)} = (\sum_{j=1}^{r} \mathbf{Y}_j{}^{(k)})/r$. With regard to the linear model Eq. (5.16), we state the following Lemma.

Lemma 1 *If* $|a_{kj}| \leq 1\overline{r}$ *(in* Eq. (5.16)*) and* $\| \mathbf{Y}_k^C - \overline{\mathbf{Y}}^{(k)} \| \leq \max(h_A^2, h_B^2)$*, for any value* $1 \leq k \leq N_c$*,* $1 \leq j \leq r$*, then the following error estimate holds*

$$\| \mathbf{Y}_k^C - \hat{\mathbf{Y}}_k \| \leq c \tag{5.18}$$

where c is a finite quantity of order $O(\max(h_A^2, h_B^2))$.

Proof. *By recalling* Eq. (5.12) *and* Eq. (5.16)*, we get*

$$\|\mathbf{Y}_k^C - \hat{\mathbf{Y}}_k\| \leq \|\mathbf{Y}_k^C - \overline{\mathbf{Y}}^{(k)}\| + O(\max(h_A^2, h_B^2)). \tag{5.19}$$

So, the conclusion can be readily achieved. ■

Lemma 1 reflects the simplest case of $\hat{\mathbf{Y}}_k$ as mean of the vectors of the kth cluster. This has practical application, because, as a first attempt, one can check through this simple case if the adopted approximation works.

5.3.3 The online stage

The online stage (Figure 5.2b) is substantially a decision-making module, of which goal is detecting the most suitable approximating function to get the optimal setting relatively to certain online measurements, without carrying out a priori any rigorous OPF calculation.

More precisely, we have the following steps:

1. Compute the discrete F-transform of the $\overline{N} \times M$ matrix of the measurements $\overline{\mathbf{X}}$, i.e., the $\overline{n} \times m$ matrix $\mathbf{F}^X$;
2. Determine the cluster to which the vector $\mathbf{F}_i^{\overline{X}}$ belongs by means of the Euclidean norm, that is, check if:

$$d(\mathbf{F}_i^{\overline{X}}, \mathbf{F}_k^C) \leq \varepsilon \tag{5.20}$$

for $1 \leq k \leq N_c$ and with $i = 1,\ldots,\overline{n}$;

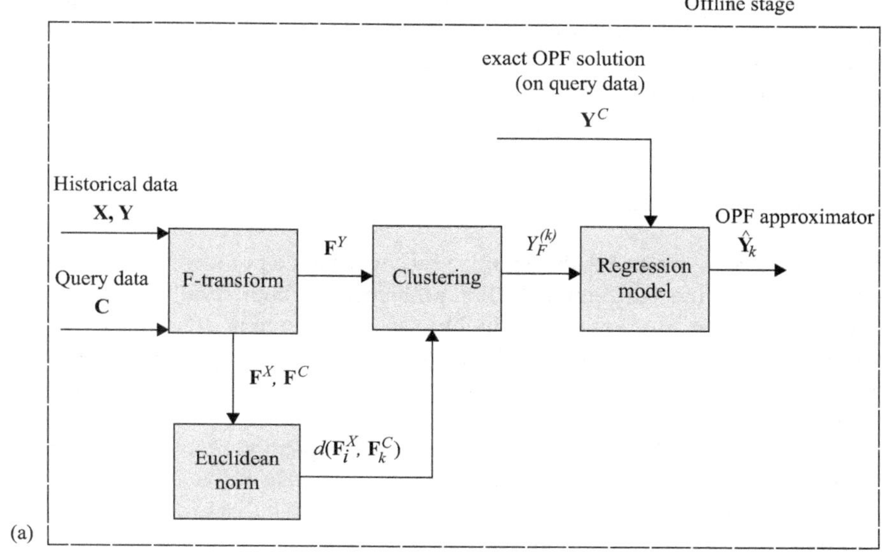

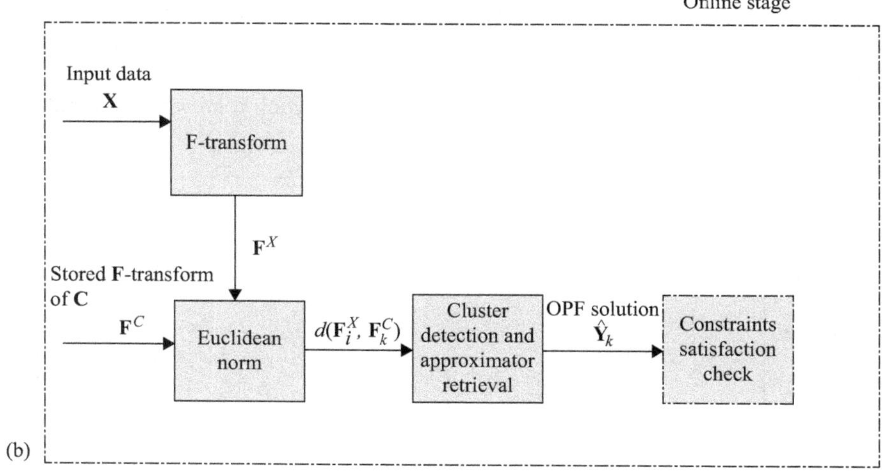

Figure 5.2 The proposed scheme: (a) the offline stage; (b) the online stage

3. Compute the approximate OPF solution by using the approximating function $\hat{\mathbf{Y}}_k$ (see Eq. (5.15) or (5.16));
4. Assess the feasibility of the approximate solution, by checking the problem constraints satisfaction, namely:

$$h(\hat{\mathbf{Y}}_k) < 0, \quad \max\left(\|g(\hat{\mathbf{Y}}_k)\|\right) < \sigma \tag{5.21}$$

where σ is a fixed allowable tolerance.

Notice that only in the case that the conditions (5.21) are not satisfied, then the approximate solution is refined by a traditional iterative algorithm aimed at rigorously solving the OPF problem. The solution so obtained, jointly with the corresponding query vector, is then used for adapting the knowledge-base by updating the cluster centers and refining the local models.

5.4 Simulation results

This section presents the results obtained by applying the proposed methodology in the task of solving conventional OPF problems for both small and large-scale power systems, in the presence of highly variable operating conditions. The first case study deals with the solution of the optimal active power dispatch problem for the IEEE 118-bus test system, which represents a portion of the Midwestern American Electric Power System composed by 118 bus, 54 generators, 64 loads, and 186 lines [43]. The control variables of this optimization problem are the active power generated by the 54 generators, while the 235 state variables are the voltage magnitude at the load buses, and the voltage angle at all buses except the slack bus. The problem objective is to minimize the total generation cost satisfying both the equality constraints, which are described by 181 power flow equations, and the inequality constraints, which are described by the allowable ranges of the voltage magnitudes at all buses, and the reactive power limits at the generation buses. The load buses are classified in 10 different customer classes, characterized by the 98 biweekly/15 min profiles depicted in Figure 5.3, which represent the input data

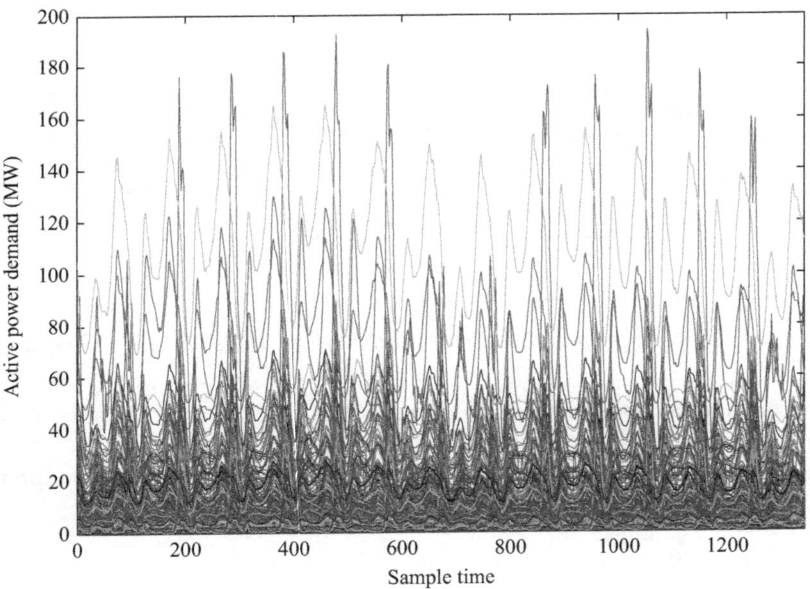

Figure 5.3 Loading profiles used for the OPF analysis of the IEEE 118-bus test system

of the optimization problem. Starting from these input profiles the optimal active power dispatch problem is solved, and the obtained results, which are depicted in Figure 5.4, were arranged in the following matrices:

- **X** and **Y**, which are composed by the first 1151 input vectors, and the corresponding solutions of the optimization problem, and whose cardinality are [1151, 98] and [1151, 235], respectively;
- The query and the validation matrices, **C** and $\mathbf{Y}^C$, which are composed by the remaining 192 input vectors and the corresponding solutions of the optimization problem, which have been adopted to verify the accuracy of the proposed methodology.

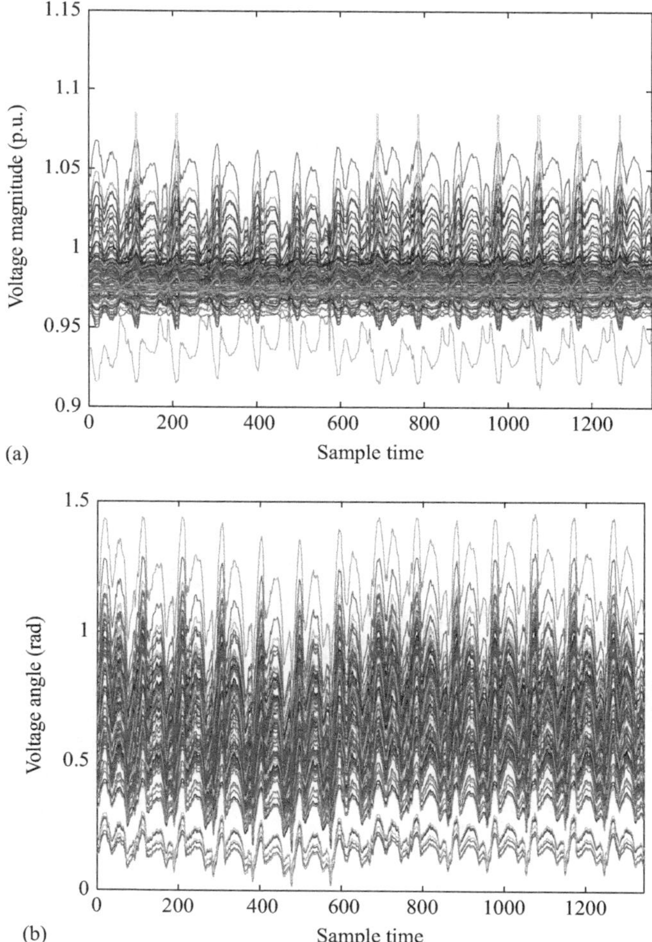

Figure 5.4 Solutions of the constrained power flow problem for the IEEE 118-bus test system: (a) bus voltage magnitude; (b) bus voltage angle

The discrete F-transform is first applied to the matrices **X** and **Y**, reducing their cardinality to [300, 25] and [300, 23], respectively, which led to a compression ratio of the order of 80%. The vectors of the transformed **X** matrix are then classified according to the proposed algorithm, with a number of 10 data clusters, and the F-transform is applied to the query matrix, leading to a reduced cardinality of [192, 25]. Finally, for each transformed query vector, the corresponding approximate OPF solution has been determined. The corresponding error surfaces and the error histograms between the identified and the rigorous solutions are reported in Figures 5.5 and 5.6, respectively. By analyzing these figures, it is possible to appreciate the high accuracy of the solutions computed by the proposed F-transform-based methodology. This is still more relevant if one considers the

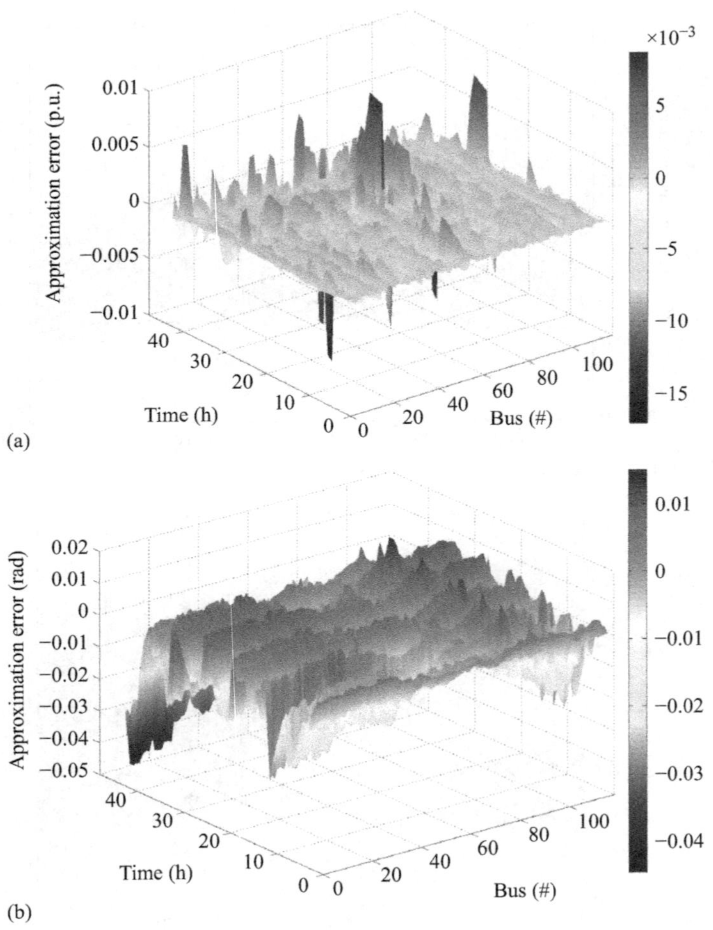

Figure 5.5 Errors between the identified and the rigorous solutions: (a) bus voltage magnitude; (b) bus voltage angle

sensible reduction of the CPU times, which are on average 95% faster with respect to those required by traditional OPF solution algorithms.

In order to further test the performance of the proposed methodology, the constrained power flow analysis of the 2383-bus Polish power system, which represents the Polish 400, 220, and 110 kV networks composed by 2383 bus, 327 generators, 2056 loads, and 2896 lines, was considered. The 4763 state variables of this optimization problem are the voltage magnitude at the load buses, and the voltage angle at all buses except the slack bus. The problem objective is to minimize the power mismatch between the computed and the fixed active and reactive bus powers, satisfying the power flow equations, and assuring that the voltage

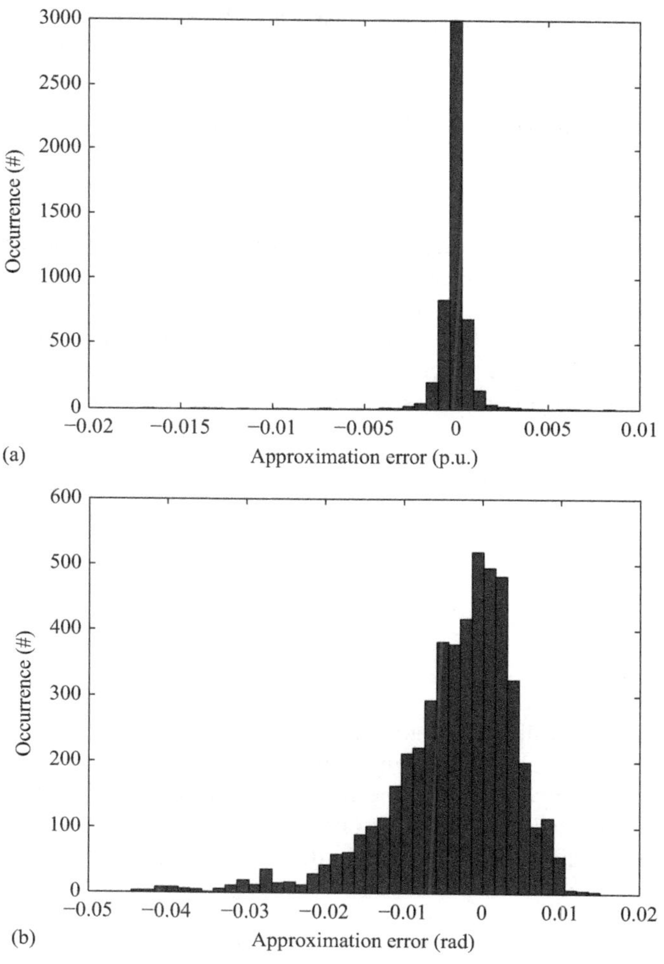

Figure 5.6 Histogram of the errors between the identified and the rigorous solutions: (a) bus voltage magnitude; (b) bus voltage angle

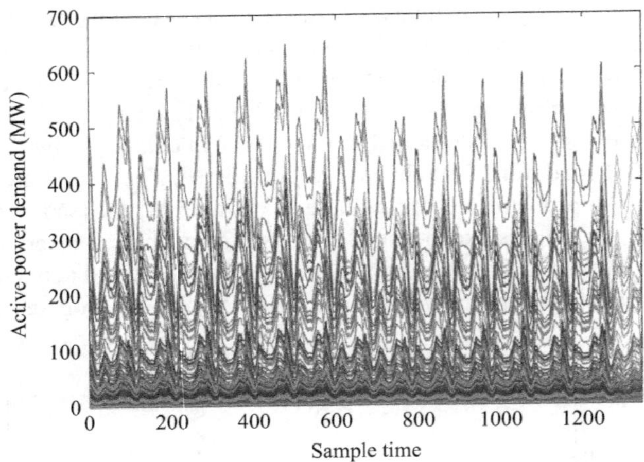

Figure 5.7 Loading profiles used for the power flow analysis of the 2382-bus Polish test system

magnitude at each buses and the reactive power generated at the generation buses are within the allowable ranges. To simulate realistic power system operation conditions, the biweekly/15 min load profiles obtained from the Australian Energy Market Operator database [44] were adopted to generate the 3644 load profiles shown in Figure 5.7. The corresponding active power generation profiles were defined for each time sample K as follows:

$$P_{G_i}(K) = \alpha_{G_i} \; r_n \sum_{j=1}^{2383} P_{D_j}(K) \quad \forall i \in [1, 327] \tag{5.22}$$

where α_{G_i} are dispatch factors, i.e.,

$$\alpha_{G_i} = \frac{P_{G_i}(0)}{\sum_{j=1}^{2383} P_{D_j}(0)} \quad \forall i \in [1, 327] \tag{5.23}$$

To obtain a more realistic scenario, these dispatch factors were perturbed with respect to their base values by multiplying them with a random noise r_n uniformly distributed in the range [0.9,1.1]. The corresponding bus voltages magnitudes and angles were computed by rigorously solving the constrained power flow problem, and the corresponding results, which are depicted in Figure 5.8, were arranged in the following matrices:

- the matrices **X** and **Y**, which are composed by the first 1151 input vectors and the corresponding solutions of the optimization problem, and whose cardinality are [1151, 3970] and [1151, 4763], respectively;

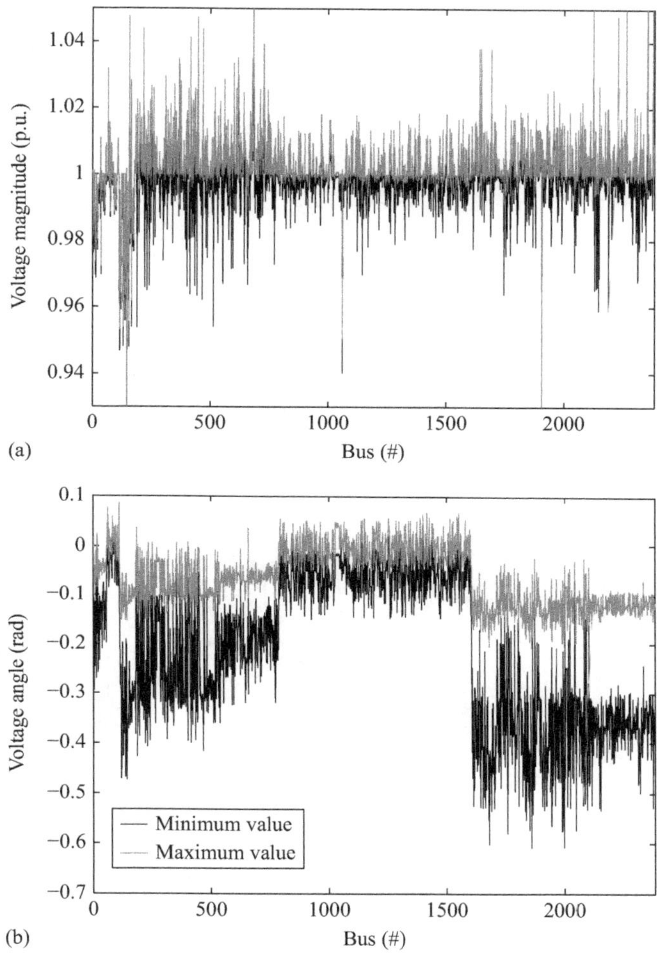

Figure 5.8 Solutions of the constrained power flow problem for the he 2382-bus Polish test system: (a) bus voltage magnitude; (b) bus voltage angle

- the query and the validation matrices, **C** and $\mathbf{Y}^C$, which are composed by the remaining 192 input vectors and the corresponding solutions, respectively.

The F-transform is first applied to the matrices **X** and **Y**, reducing their cardinality to [400, 200] and [400, 4763], respectively, which led to a compression ratio of about 80%. The vectors of the transformed **X** matrix are then classified according to the proposed offline algorithm, with a number of 20 data clusters, and the F-transform is applied to the query matrix, leading to a reduced cardinality of [192, 200]. Finally, for each transformed query vector, the corresponding approximate power flow

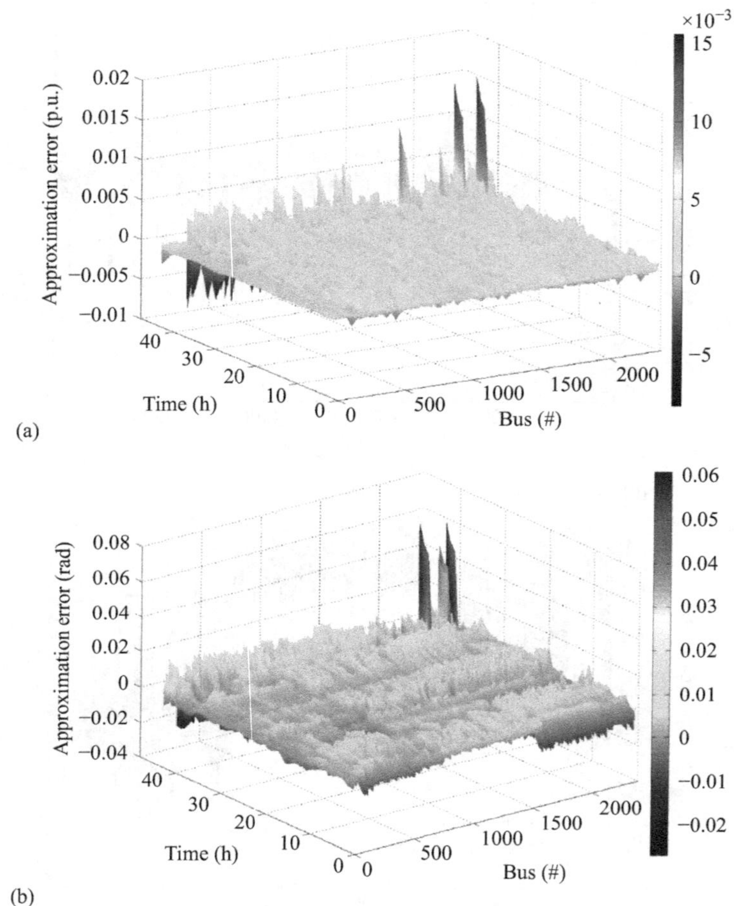

Figure 5.9 Errors between the identified and the rigorous solutions: (a) bus voltage magnitude; (b) bus voltage angle

solution has been determined. The obtained error surfaces and the error histograms between the identified and the rigorous solutions are reported in Figures 5.9 and 5.10, respectively. By analyzing these figures, it is possible to confirm the effectiveness of the proposed methodology in the task of approximating the solutions of complex and large-scale optimization problems, by sensibly reducing the CPU times, which are on average 90% faster with respect to those required by the rigorous solution algorithms.

Finally, we wish to point out that the results herein discussed have been obtained by means of a nonlinear regression model, even if the initial attempt by using the mean value through (5.16) was good enough (maximum error of order 10^{-2} for both cases on the voltage magnitude).

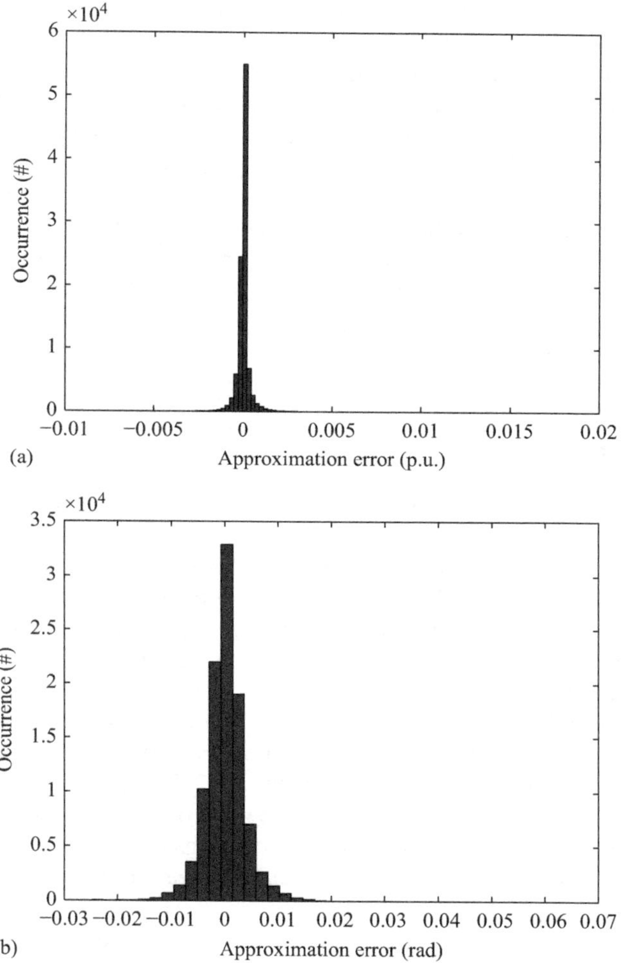

Figure 5.10 Histogram of the errors between the identified and the rigorous solutions: (a) bus voltage magnitude; (b) bus voltage angle

5.5 Conclusive remarks

In this chapter, a computational framework based on F-transform is proposed for reducing the complexity of smart grids optimization. Due to the availability of a large quantity of historical data, the adoption of the F-transform is made possible to determine an approximate OPF solution, with a sensible reduction of the computational burden of the OPF problem, in addition to the reduction of the storage occupancy. The proposed approach can be considered as part of a "big data" processing engine for smart grid optimization, in which valuable information are extracted from large volume of historical data, and delivered at the right time and to the right smart grid operator.

The numerical results obtained for realistic power systems under various operating scenario demonstrate that the overall complexity of the OPF problems could be sensibly reduced by deploying the proposed methodology, especially in the presence of correlated state variables. In particular, it was observed that the approximation accuracy and the computational burdens observed during the experiments were strictly influenced by the number of clusters identified to classify the smart grid state vectors. Therefore, formal methods aimed at defining a proper tradeoff between the solutions accuracy and the algorithm complexity would be necessary for a comprehensive deployment of the proposed approach. This topic is currently under investigation.

References

[1] R.L. King, "Information services for smart grids", *Proceedings of the IEEE Power and Energy Society General Meeting – Conversion and Delivery of Electrical Energy in the 21st Century*, Pittsburgh, PA, 20–24 July 2008, pp. 1–5.

[2] L. Madani, S. Vahid, R.L. King, Roger, "Strategies and applications to meet grid challenges and enhance power system performance", *Proceedings of the IEEE Power Engineering Society General Meeting*, Tampa, FL, 24–28 June 2007, pp. 1–9.

[3] J. Baek, Q. Vu, J. Liu, X Huang, Y. Xiang, "A secure cloud computing based framework for big data information management of smart grid", *IEEE Trans Cloud Comp*, vol. to appear, pp. 1–12, 2015.

[4] M. Kezunovic, X. Le, S. Grijalva, "The role of big data in improving power system operation and protection", *Proceedings of the 2013 IREP Symposium on Bulk Power System Dynamics and Control,* Rethymno, 25–30 August 2013, pp. 1–9.

[5] A. Gomez-Exposito, A.J. Conejo, C.A. Cañizares, *Electric Energy Systems: Analysis and Operation*, 2009, CRC Press, Boca Raton, FL.

[6] P.J. Lagace, "Power flow methods for improving convergence", *Proceedings of the 38th Annual Conference on IEEE Industrial Electronics Society*, Montreal, QC, 25–28 October 2012, pp. 1387–1392.

[7] A. Sousa, G.L. Torres, "Globally convergent optimal power flow by trust-region interior-point methods", *Proceedings of the IEEE Lausanne Power Tech*, Lausanne, 1–5 July 2007, pp. 1386–1391.

[8] A. Gomez-Exposito, C. Gamez-Quiles, "Factorized load flow", *IEEE Transactions on Power Systems*, vol. 28, no. 4, pp. 4607–4614, 2013.

[9] F.J. Nogales, F.J. Prieto, A.J. Conejo, "A decomposition methodology applied to the multi-area optimal power flow problem", *Annals of Operations Research*, vol. 20, no. 4, pp. 99–116, 2003.

[10] K.P. Wong, J. Yuryevich, A. Li, "Evolutionary-programming-based load flow algorithm for systems containing unified power flow controllers",

IET Generation, Transmission Distribution, vol. 150, no. 4, pp. 441–446, 2003.

[11] L.L. Lai, J.T. Ma, R. Yokoyama, M. Zhao, "Improved genetic algorithms for optimal power flow under both normal and contingent operation states", *International Journal of Electrical Power Energy Systems*, vol. 19, no. 5, pp. 287–292, 1997.

[12] P.A. Pajan, V.L. Paucar, "Fuzzy power flow: considerations and application to the planning and operation of a real power system", *Proceedings of the International Conference On Power System Technology*, Lima, Peru, 13–17 October 2002, pp. 433–437.

[13] S. Kumar, D.K. Chaturvedi, "Optimal power flow solution using fuzzy evolutionary and swarm optimization", *International Journal on Electrical Power & Energy Systems*, vol. 47, no. 1, pp. 416–423, 2013.

[14] H. H. Muller, M.J. Rider, C.A. Castro, V.L. Paucar, "Power flow model based on artificial neural networks", *Proceedings of the IEEE Power Technology*, St. Petersburg, 27–30 June 2005, pp. 1–6.

[15] V.J. Gutierrez-Martinez, C.A. Cañizares, C.R. Fuerte-Esquivel, A. Pizano-Martinez, X. Gu, "Neural-network security-boundary constrained optimal power flow", *IEEE Transactions on Power Systems*, vol. 26, no. 1, pp. 63–72, 2011.

[16] Q. Morante, N. Ranaldo, A. Vaccaro, E. Zimeo, "Pervasive grid for large-scale power systems contingency analysis", *IEEE Transactions on Industrial Informatics*, vol. 2, no. 3, pp. 165–175, 2006.

[17] S.-Y. Lin, J.-F. Chen, "Distributed optimal power flow for smart grid transmission system with renewable energy sources", *Energy*, vol. 56, no. 1, pp. 184–192, 2013.

[18] D. Phan, J. Kalagnanam, "Some efficient optimization methods for solving the security-constrained optimal power flow problem", *IEEE Transactions on Power Systems*, vol. 29, no. 2, pp. 863–872, 2014.

[19] L. Platbrood, F. Capitanescu, C. Merckx, H. Crisciu, L. Wehenkel, "A generic approach for solving nonlinear-discrete security-constrained optimal power flow problems in large-scale systems", *IEEE Transactions on Power Systems*, vol. 29, no. 3, pp. 1194–1203, 2014.

[20] D.K. Molzahn, I.A. Hiskens, "Sparsity-exploiting moment-based relaxations of the optimal power flow problem", *IEEE Transactions on Power Systems*, vol. 30, no. 6, pp. 3168–3180, 2015.

[21] R. Yang, G. Hug-Glanzmann, "Regression-based control of thyristor-controlled series compensators for optimal usage of transmission capacity", *IET Generation Transmission & Distribution*, vol. 8, no. 8, pp. 1444–1452, 2014.

[22] I. Perfilieva, "Fuzzy transforms: theory and applications", *Fuzzy Sets Systems*, vol. 157, pp. 993–1023, 2006.

[23] F. Di Martino, V. Loia, S. Sessa, "Fuzzy transforms method and attribute dependency in data analysis", *Information Sciences*, vol. 180, pp. 493–505, 2010.

[24] F. Di Martino, V. Loia, I. Perfilieva, S. Sessa, "An image coding/decoding method based on direct and inverse fuzzy transforms", *International Journal of Approximate Reasoning*, vol. 48, pp. 110–131, 2008.

[25] F. Di Martino, V. Loia, I. Perfilieva, S. Sessa, "Fuzzy transform for coding/decoding images: a short description of methods and techniques", *Studies in Fuzziness and Soft Computing*, vol. 298, pp. 139–146, 2013.

[26] P. Vlasanek, I. Perfilieva, "Influence of various types of basic functions on image reconstruction using F-transform", *Advances in Intelligent Systems Research*, vol. 32, pp. 497–502, 2013.

[27] P. Hurtik, I. Perfilieva, "Image compression methodology based on fuzzy transform using block similarity", *Advances in Intelligent Systems Research*, vol. 32, pp. 521–526, 2013.

[28] F. Di Martino, V. Loia, S. Sessa, "Fuzzy transforms method and attribute dependency in data analysis", *Information Sciences*, vol. 180, pp. 493–505, 2010.

[29] J.K.I. Tomanova, "Hidden functional dependencies found by the technique of F-transform", *Advances in Intelligent Systems Research*, vol. 32, pp. 662–668, 2013.

[30] I. Perfilieva, V. Novak, V. Pavliska, A. Dvorak, M. Stepnicka, "Analysis and prediction of time series using fuzzy transform", *Proceedings of the IEEE World Congress Computational Intelligence*, Hong Kong, 1–8 June 2008, pp. 3875–3879.

[31] I. Perfilieva, N. Yarushkina, T. Afanasieva, A. Romanov, "Time series analysis using soft computing methods", *International Journal of General Systems*, vol. 42, no. 6, pp. 687–705, 2013.

[32] V. Novak, V. Pavliska, I. Perfilieva, M. Stepnicka, "F-transform and fuzzy natural logic in time series analysis", *Advances in Intelligent Systems Research*, vol. 32, pp. 40–47, 2013.

[33] M. Pirnia, C. Cañizares, A. Claudio, K. Bhattacharya, "Revisiting the power flow problem based on a mixed complementarity formulation approach", *IET Generation, Transmission & Distribution*, vol. 7, no. 11, pp. 1194–1201, 2013.

[34] R.R. Shoults, D.T. Sun, "Optimal power flow based upon PQ decomposition", *IEEE Transactions on Power Systems*, vol. 2, pp. 397–405, 1982.

[35] A. Momoh, A. James, "A generalized quadratic-based model for optimal power flow", *Proceedings of the IEEE International Conference Systems, Man and Cybernetics*, Cambridge, MA, 14–17 November 1989, pp. 261–271.

[36] R.C. Burchett, H.H. Happ, K.A. Wirgau, "Large scale optimal power flow", *IEEE Transactions on Power Systems*, vol. 10, pp. 3722–3732, 1982.

[37] K.S. Pandya, S.K. Joshi, "A survey of optimal power flow methods", *Journal of Theoretical and Applied Information Technology*, vol. 4, no. 5, pp. 450–458, 2008.

[38] C.L. Stimmel, *Big Data Analytics Strategies for the Smart Grids*, CRC Press Taylor and Francis Group, 2015, Boca Raton, FL.

[39] P.M. Devie, S.Kalyani, "An optimization framework for cloud-based data management model in smart grids", *International Journal of Research in Engineering and Technology*, vol. 4, no. 2, pp. 751–758, 2015.

[40] L. Liu, Z. Han, "Multi-block ADMM for big data optimization in smart grids", *Proceedings of the International Conference on Computing, Networking and Communications (ICNC 2015)*, Anaheim, California, USA, February 16–19, 2015, pp. 1–6.

[41] E. Dall'Anese, H. Zhu, G.B. Giannakis, "Distributed optimal power flow for smart microgrids", *IEEE Transactions on Smart Grid*, vol. 4, no. 3, pp. 1464–1475, 2013.

[42] M. Stepnicka, *Fuzzy Transform and its Applications to Problems in Engineering Practice*, PhD thesis, University of Ostrava, 2007.

[43] R.D. Christie, Power systems test case archive, available online at http://www.ee.washington.edu/research/pstca, last access May 2015.

[44] Australian Energy Market Operator, Energy Market Data, available on line at http://www.aemo.com.au/Electricity/Data, last access May 2015.

Chapter 6

False data injection attacks and countermeasures for wide area measurement system

Junbo Zhao[1,2] and Massimo La Scala[3]

6.1 Introduction

State estimation (SE), which is the backbone of the modern energy management system (EMS), is extremely important for ensuring power system reliable operation and control. It provides system accurate and continuously updated snapshots of the real-time states, which enable EMS to perform various important controls and planning tasks such as optimal power flows, voltage stability study, and contingency analysis [1]. However, with the development of smart grids and installation of the wide area measurement system (WAMS), the power systems are more vulnerable to information failures and malicious attacks. One of them is the cyber-attacks against SE, which can mislead the system controls, possibly resulting in catastrophic large geographical blackouts [2].

After the introduction of the false data injection attack (FDIA) into power grids [2], it has been the object of new interests and investigations among researchers and utilities because of the potential attack risks due to the increasing number of links to public networks and the web-based applications in the power industry, etc. The FDIA can successfully bypass the conventional normalized measurement residual-based bad data (BD) detection, thus causing serious threats to system operation and control. To date, three kinds of FDIAs have been proposed, i.e., state attack [2–4], topology attack [5], and load redistribution attack [6]. In the state attack case, the adversary could introduce arbitrary perturbations into the system SE results by altering the measurement values of a set of meters. In the topology attack, the adversary aims to compromise a certain number of meters and break circuit switches to mislead the operator with the incorrect system topology without being detected. In the load redistribution attack scenario, power injection measurements of the load buses and line power flow measurements were attacked and used to change the power flow

[1]School of Electrical Engineering, Southwest Jiaotong University, Chengdu, 610031 China
[2]Bradley Department of Electrical Computer Engineering, Virginia Polytechnic Institute and State University, Northern Virginia Center, Falls Church, Virginia 22043, USA
[3]Dipartimento di Elettrotecnica ed Elettronica (DEE), Politecnico di Bari, 70125 Bari, Italy

distributions, i.e., to increase loads at some buses and to reduce loads at other buses without changing the total loads. Except for the FDIAs on simplified linear measurement model-based SE, i.e., DC FDIAs, a few papers have explored and investigated the FDIAs on practical nonlinear measurement model-based SE, i.e., AC FDIAs. In Reference 7, the vulnerability of FDIA on AC SE and the minimal required manipulation of measurement values was analyzed and discussed. In Reference 8, a special model for FDIA against nonlinear SE was presented and two types of attacks: perfect and imperfect attacks were defined. Simulations for different single state variable attacks verify its effectiveness. In Reference 9, the effects of perfect attacks on real-time electric market were investigated. While, in Reference 10, a forecasting-aided FDIA on AC SE was proposed, where the forecasted measurements were used for constructing the attack vector and its associated forecasting error on affecting the successfully attack possibility was analyzed.

To detect and mitigate the FDIAs, a number of methods have been proposed [11–16]. Two security indices based on the analysis of the sparsity of attack vector and attack vector magnitude were proposed [11]. The least effort needed to launch the FDIA while avoiding detections by the control center was also discussed. In Reference 12, a greedy algorithm-based secure phasor measurement unit (PMU) placement method was proposed to defend against FDIA. In Reference 13, the FDIA was formulated as a matrix separation problem while the nuclear norm minimization and low rank matrix factorization methods were used as detection metrics. In Reference 14, known perturbations were applied to the system and then the system was "probed" for any unexpected responses; however, the random known perturbations, i.e., topology, transformer taps, etc., cannot guarantee the complete elimination of the FDIA possibility. For example, it was shown in Reference 15 that the FDIA was still successful if the attacker has imperfect but structured topological information of the system. An alternative group of the FDIA defense methods [16], [17] aimed to add protections on many measurements so that the adversary could not get enough measurements to launch attacks. For instance, in Reference 12, a specific selected measurement-based strategy was adopted against the attacks. The selected measurements were the minimum number of measurements needed to ensure system observable.

In this chapter, the authors propose to extend the existing FDIAs on DC or AC model-based SE to a more generalized framework, where both perfect and imperfect attacks are included. To detect or mitigate FDIAs, a measurement consistency check-based index is proposed. This index is defined by the normalized measurement difference between received measurements and the interpolated measurements that are calculated through a small number of selected secure alternative PMU measurements. This index works well even when BD and FDIAs occur simultaneously.

The organization of this chapter is as follows: Section 6.2 presents a brief review of AC and DC model-based SE methods, and the associated BD detection and processing techniques. The framework for FDIAs on DC model-based SE is given in Section 6.3, while Section 6.4 presents the framework for FDIAs on AC model-based SE. In Section 6.5, a novel measurement consistency check-based FDIAs detection method is presented, followed by the simulation results in Section 6.6. Finally, Section 6.7 concludes this chapter.

6.2 Preliminaries of SE

6.2.1 Nonlinear SE

For an n-buses power system using AC power flow model, the relationship between the measurement vector $\mathbf{z}$ (i.e., active and reactive power flows and injections as well as voltage magnitudes) and state vector $\mathbf{x}$ (i.e., voltage magnitudes and angles) can be described as follows:

$$\mathbf{z} = \mathbf{h}(\mathbf{x}) + \mathbf{v} \tag{6.1}$$

where $\mathbf{h}(\cdot)$ is the nonlinear power flow function vector, $\mathbf{v}$ is the random measurement error vector and is usually assumed to be normally distributed with zero mean and covariance matrix $\mathbf{R}$, i.e., $\mathbf{v} \sim N(\mathbf{0}, \mathbf{R})$. The widely used iterative Gauss–Newton method [18] is adopted to solve the following objective function

$$\hat{\mathbf{x}} = \arg \min_{\mathbf{x}} [\mathbf{z} - \mathbf{h}(\mathbf{x})]^T \mathbf{R}^{-1} [\mathbf{z} - \mathbf{h}(\mathbf{x})] \tag{6.2}$$

with the most basic form of iteration updates

$$\hat{\mathbf{x}}_{k+1} = \hat{\mathbf{x}}_k + \delta_k, k = 1, 2, \ldots \tag{6.3}$$

where the iterative descent δ_k with a chosen initializer $\mathbf{x}_0$ can be expressed as:

$$\delta_k = \left(\mathbf{J}^T \mathbf{R}^{-1} \mathbf{J}\right)^{-1} \mathbf{J}^T \mathbf{R}^{-1} \left[\mathbf{z} - \mathbf{h}(\mathbf{x}_k)\right] \tag{6.4}$$

where $\mathbf{J} = \partial \mathbf{h} / \partial \mathbf{x}|_{\mathbf{x}=\mathbf{x}_k}$ is the Jacobian matrix. The algorithm converges once δ_k becomes smaller than a prespecified tolerance threshold, i.e., 10^{-2}.

6.2.2 DC state estimation

The nonlinear model in (6.1) can be simplified to the following linear model by assuming the voltage magnitudes at every node equal to one, the angle differences between two buses are very small and the shunt susceptance and series resistances in the transmission lines are ignored [18, 19]:

$$\mathbf{z} = \mathbf{H}\mathbf{x} + \mathbf{v} \tag{6.5}$$

where $\mathbf{H}$ represents the linear dependencies between measurement and state variables, which is determined only by the system topology and line parameters. Thus, the objective function for this linear regression problem is:

$$\min_{\mathbf{x}} [\mathbf{z} - \mathbf{H}\mathbf{x}]^T \mathbf{R}^{-1} [\mathbf{z} - \mathbf{H}\mathbf{x}] \tag{6.6}$$

Using the weighted least squares (WLS) estimation method, the state vector $\hat{\mathbf{x}}$ can be estimated as:

$$\hat{\mathbf{x}} = \left(\mathbf{H}^T \mathbf{R}^{-1} \mathbf{H}\right)^{-1} \mathbf{H}^T \mathbf{R}^{-1} \mathbf{z} \tag{6.7}$$

Even though the WLS method is widely used to solve the power system SE problem with quadratic convergence rate, it is sensitive to the BD due to its nonrobustness [20, 21]. Therefore, the BD detection and processing methods are proposed.

6.2.3 Bad data detection

After the SE, the widely recognized L_2-norm measurement residual-based detector or the largest normalized residual-based (LNR) [18] detector is adopted to detect and eliminate the BD. The LNR method uses the following condition to decide if there is any BD in the measurement set. If the inequality

$$\|\mathbf{r}\| = \|\mathbf{z} - \mathbf{h}(\hat{\mathbf{x}})\| \leq \tau \qquad \text{or} \qquad \|\mathbf{r}\| = \|\mathbf{z} - \mathbf{H}\hat{\mathbf{x}}\| \leq \tau \tag{6.8}$$

is violated, then there is at least one suspicious measurement. The τ is the threshold and is determined by the confidence level.

The suspicious measurements will be removed one by one and the nonlinear SE problem will be solved iteratively until all the suspicious measurements are removed and the final state estimates are obtained. If BD is not properly detected or eliminated, the SE results will be biased leading to unreliable network operation and optimization.

6.3 Framework of FDIA on DC SE

6.3.1 Generic linear measurement model

In the literature, the approximately simplified and linearized DC power flow model derived from the complex nonlinear power flow equations is widely used for the FDIA construction. This pure DC model-based FDIA is neither accurate nor general for the following reasons: from the perspective of the operator, inclusion of more accurate PMU measurements into SE will generate more accurate estimation results [22]. Then, the pure supervisory control and data acquisition (SCADA) measurement-based SE will be slightly modified because with the increasing installation of PMU devices, part of the Jacobian matrix will be exactly linear for the PMU observable area or even the whole Jacobian matrix will be exactly linear if the number of PMUs is enough for ensuring the entire system observable. In this situation, the linearization errors are reduced. From the adversaries' point, if the pure approximate DC model is still used for the FDIA construction, larger deviations will be produced, thus resulting in being detected with higher probability by the control center since the measurement model used by the control center has been modified. In this section, a more general linear measurement model that can handle both conventional SCADA and PMU measurements is derived. Then, the general FDIA on this model is presented and discussed.

For any linear measurement model, the relationship between the measurement vector $\mathbf{z}$ and the state vector $\mathbf{x}$ is:

$$\mathbf{z} = \mathbf{H}\mathbf{x} + \mathbf{v} \tag{6.9}$$

where $\mathbf{H}$ is a matrix that represents the linear relationship between the measurements and the states. For PMU full observable systems, the elements of $\mathbf{H}$ are constituted by the system conductance and susceptance without any linearization error, while for PMU partial observable systems, $\mathbf{H} = \left[\mathbf{H}_c^T\ \mathbf{H}_p^T\right]^T$, where $\mathbf{H}_c$ is the approximate DC model related part for the PMU unobservable area, and $\mathbf{H}_p$ is the accurate linear model for the PMU observable area. If no PMUs are installed in the system, $\mathbf{H} = \mathbf{H}_c$.

6.3.2 Generic FDIA on the proposed linear measurement model

In this subsection, the perfect FDIA [2] will be imposed on the proposed linear measurement model. Then, the generic FDIA on the proposed linear measurement model will be proposed and analyzed.

1. Perfect FDIA on the proposed linear measurement model

   In this case, we assume that the hacker can get the same matrix $\mathbf{H}$ as the control center to construct the attack vector $\mathbf{a}$ as:

   $$\mathbf{a} = \mathbf{Hc} \tag{6.10}$$

   where $\mathbf{c}$ is the nonzero column vector that represents the attack magnitude on the estimated state vector. Therefore, if there is no BD in the measurement set, i.e., $\|\mathbf{z} - \mathbf{H}\hat{\mathbf{x}}\| \leq \tau$, the injected false data cannot be detected by the residual analysis-based BD detection algorithm, because:

   $$\begin{aligned} \|\mathbf{r}\| &= \|\mathbf{z} + \mathbf{a} - \mathbf{H}(\hat{\mathbf{x}} + \mathbf{c})\| = \|\mathbf{z} - \mathbf{H}\hat{\mathbf{x}} + \mathbf{a} - \mathbf{Hc}\| \\ &= \|\mathbf{z} - \mathbf{H}\hat{\mathbf{x}}\| \leq \tau \end{aligned} \tag{6.11}$$

   The estimated state vector will be:

   $$\hat{\mathbf{x}}_a = \left(\mathbf{H}^T\mathbf{R}^{-1}\mathbf{H}\right)^{-1}\mathbf{H}^T\mathbf{R}^{-1}(\mathbf{z} + \mathbf{a}) = \hat{\mathbf{x}} + \mathbf{c} \tag{6.12}$$

   Remark: *Actually, the assumption that hackers could acquire perfect system configuration information, i.e., Jacobian matrix* **H** *without biases, is not practical for real-power systems. This is because the attacker is lack of real-time knowledge with respect to the status of various grid elements such as the position of circuit breaker switches and transformer tap changers, and also because he/she is restricted to get access to many grid facilities (e.g., the hacker may not know the new installed PMU devices while the control center can get access to them). Thus, it is impossible for the hacker to get exactly the same Jacobian matrix* **H** *as the control center. In other words, the* **H** *that the hacker gets has bias, which would increase the risk of being detected by the control center since the FDIA is not perfect any more. Therefore, how to find the general FDIA model (including perfect and imperfect FDIAs) while maintaining as low probability to be detected by the control center as possible should be investigated.*

2. Generic FDIA on the proposed linear measurement model

   In this case, the matrix $\mathbf{H}$ the hacker obtains has bias $\boldsymbol{\varepsilon}$, i.e., $\mathbf{H} \leftarrow \mathbf{H} + \boldsymbol{\varepsilon}$, where the bias is due to the imperfect knowledge of the system information. Thus, the attack vector constructed by the adversary in this scenario will be:

$$\mathbf{a} = (\mathbf{H} + \boldsymbol{\varepsilon})\mathbf{c} = \mathbf{Hc} + \boldsymbol{\varepsilon}\mathbf{c} \tag{6.13}$$

The estimated state vector being attacked is:

$$\begin{aligned} \hat{\mathbf{x}}_a &= \left(\mathbf{H}^T\mathbf{R}^{-1}\mathbf{H}\right)^{-1}\mathbf{H}^T\mathbf{R}^{-1}(\mathbf{z} + \mathbf{a}) \\ &= \left(\mathbf{H}^T\mathbf{R}^{-1}\mathbf{H}\right)^{-1}\mathbf{H}^T\mathbf{R}^{-1}(\mathbf{z} + \mathbf{Hc} + \boldsymbol{\varepsilon}\mathbf{c}) \\ &= \hat{\mathbf{x}} + \mathbf{c} + \left(\mathbf{H}^T\mathbf{R}^{-1}\mathbf{H}\right)^{-1}\mathbf{H}^T\mathbf{R}^{-1}\boldsymbol{\varepsilon}\mathbf{c} \end{aligned} \tag{6.14}$$

By comparing (6.12) and (6.14), it is observed that the intended attack magnitude $\mathbf{c}$ on the state vector has changed to $\overline{\mathbf{c}} = \mathbf{c} + \left(\mathbf{H}^T\mathbf{R}^{-1}\mathbf{H}\right)^{-1}\mathbf{H}^T\mathbf{R}^{-1}\boldsymbol{\varepsilon}\mathbf{c}$ due to the imperfect knowledge of system matrix $\mathbf{H}$. The residual can be obtained as:

$$\begin{aligned} \mathbf{r}_a &= \mathbf{z} + \mathbf{a} - \mathbf{H}(\hat{\mathbf{x}} + \overline{\mathbf{c}}) = \mathbf{r} + \mathbf{a} - \mathbf{H}\overline{\mathbf{c}} \\ &= \mathbf{r} + \boldsymbol{\varepsilon}\mathbf{c} + \mathbf{H}(\mathbf{c} - \overline{\mathbf{c}}) \\ &= \mathbf{r} + \boldsymbol{\varepsilon}\mathbf{c} - \left(\mathbf{H}^T\mathbf{R}^{-1}\mathbf{H}\right)^{-1}\mathbf{H}^T\mathbf{R}^{-1}\boldsymbol{\varepsilon}\mathbf{c} \\ &= \mathbf{r} + (\mathbf{I} - \mathbf{M})\boldsymbol{\varepsilon}\mathbf{c} = \mathbf{r} + \mathbf{S}\boldsymbol{\varepsilon}\mathbf{c} \end{aligned} \tag{6.15}$$

where $\mathbf{M} = \left(\mathbf{H}^T\mathbf{R}^{-1}\mathbf{H}\right)^{-1}\mathbf{H}^T\mathbf{R}^{-1}$; $\mathbf{S} = \mathbf{I} - \mathbf{M}$ is the sensitivity matrix [18]; $\mathbf{I}$ is the identity matrix.

Proposition 1. *Suppose the original measurement* $\mathbf{z}$ *can bypass the* L_2*-norm measurement residual-based BD detection. The malicious measurement* $\mathbf{z} + \mathbf{a}$ *can also pass this detector as long as the following condition holds, i.e.,*

$$\lambda = \|\mathbf{a} - \mathbf{H}\overline{\mathbf{c}}\| = \|(\mathbf{I}-\mathbf{M})\boldsymbol{\varepsilon}\mathbf{c}\| = \|\mathbf{I}-\mathbf{M}\|\|\boldsymbol{\varepsilon}\mathbf{c}\| \leq \tau - \|\mathbf{z} - \mathbf{H}\hat{\mathbf{x}}\|$$

where $\mathbf{a}$ *is the sparse attack vector,* $\mathbf{c}$ *is the nonzero column vector,* λ *is the error upper bound of the attack vector, and* τ *is the BD detection threshold.*

Proof *since* $\mathbf{z}$ *can bypass the* L_2*-norm measurement residual-based BD detection,* $\|\mathbf{z} - \mathbf{H}\hat{\mathbf{x}}\| \leq \tau$ *holds. Thus, the* L_2*-norm residual of the attacked measurements is:*

$$\begin{aligned} \|\mathbf{r}_a\| &= \|\mathbf{r} + (\mathbf{I} - \mathbf{M})\boldsymbol{\varepsilon}\mathbf{c}\| \\ &\leq \|\mathbf{r}\| + \|(\mathbf{I} - \mathbf{M})\boldsymbol{\varepsilon}\mathbf{c}\| = \|\mathbf{r}\| + \|\mathbf{I}-\mathbf{M}\|\|\boldsymbol{\varepsilon}\mathbf{c}\| = \|\mathbf{r}\| + \lambda \\ &\leq \|\mathbf{r}\| + \tau - \|\mathbf{z} - \mathbf{H}\hat{\mathbf{x}}\| = \tau \end{aligned} \tag{6.16}$$

where the condition $\lambda \leq \tau - \|\mathbf{z} - \mathbf{H}\hat{\mathbf{x}}\|$ *is used from row* 2 *to* 3 *in* (6.16)*; the definition of the residual* $\|\mathbf{r}\| = \|\mathbf{z} - \mathbf{H}\hat{\mathbf{x}}\|$ *when no attack occurs is used in the row* 3 *of* (6.16). *Thus, from* (6.16), *we can conclude the attack cannot be detected by the residual-based BD detection method.*

Remark: *Actually, the upper bound of λ represents the attack magnitude and degree of imperfect knowledge of the system information. Once the error of the matrix* **H** *due to the imperfect knowledge of the system is large, the attack magnitude imposed on the state by the hacker is restricted a lot in order not to be detected by the control center. Otherwise, the hacker can have a relatively large freedom of choosing the attack magnitude within the upper error bound.*

Remark: *According to the definition of critical measurement* [18], *the zero diagonal(s) of* **S** *is (are) defined as critical measurement(s). It is interesting to notice that as long as the attacks are imposed on the critical measurements,*

$$\lambda = \|\mathbf{a} - \mathbf{H}\bar{\mathbf{c}}\| = \|(\mathbf{I}-\mathbf{M})\boldsymbol{\varepsilon}\mathbf{c}\| = \|\mathbf{I}-\mathbf{M}\|\|\boldsymbol{\varepsilon}\mathbf{c}\| = 0 \le \tau - \|\mathbf{z} - \mathbf{H}\hat{\mathbf{x}}\| \quad (6.17)$$

always holds irrespective of the attack magnitudes. This on the other hand means that

$$\begin{aligned} \|\mathbf{r}_a\| &= \|\mathbf{r} + (\mathbf{I} - \mathbf{M})\boldsymbol{\varepsilon}\mathbf{c}\| \\ &\le \|\mathbf{r}\| + \|(\mathbf{I} - \mathbf{M})\boldsymbol{\varepsilon}\mathbf{c}\| = \|\mathbf{r}\| + \|\mathbf{I} - \mathbf{M}\|\|\boldsymbol{\varepsilon}\mathbf{c}\| = \|\mathbf{r}\| \\ &\le \tau - \|\mathbf{z} - \mathbf{H}\hat{\mathbf{x}}\| \end{aligned} \quad (6.18)$$

is always satisfied. Therefore, the attack on critical measurements is not detectable irrespective of the attack magnitude imposed on the state.

6.4 Framework of FDIA on AC SE

In this section, the perfect FDIA on AC SE will be first introduced and analyzed, followed by the analysis of the practical FDIA and the upper bound of the attack magnitude.

6.4.1 Perfect FDIA on AC SE

The common goal of the existing FDIAs, i.e., state attacks, topology attacks and load redistribution attacks, is to change or affect the SE results, leading to system wrong operations or controls while avoiding detections. Let us define $\hat{\mathbf{x}}$ the true estimated state vector by the control center, **a** and **c** the changes in the attacked measurements and state variables, respectively. Assume that the hacker is powerful enough so that he/she can get access to the measurements that he needs through the cyber-attacks. Therefore, he/she is able to get the same estimates as the control center. In this scenario, the attack vector is constructed as $\mathbf{a} = \mathbf{h}(\hat{\mathbf{x}} + \mathbf{c}) - \mathbf{h}(\hat{\mathbf{x}})$, where $\mathbf{h}(\cdot)$ is the nonlinear measurement function.

Proposition 2. *Suppose the original measurement* $\mathbf{z}$ *can bypass the* L_2*-norm measurement residual-based BD detection. The malicious measurement* $\mathbf{z} + \mathbf{a}$ *can also pass this detector.*

Proof *since* $\mathbf{z}$ *can bypass the* L_2*-norm measurement residual-based BD detection,* $\|\mathbf{r}\| = \|\mathbf{z} - \mathbf{h}(\hat{\mathbf{x}})\| \leq \tau$ *holds. Thus, the* L_2*-norm residual of the attacked measurements is:*

$$\begin{aligned}\|\mathbf{r}_a\| &= \|\mathbf{z} + \mathbf{a} - \mathbf{h}(\hat{\mathbf{x}} + \overline{\mathbf{c}})\| = \|\mathbf{z} + \mathbf{a} - \mathbf{h}(\hat{\mathbf{x}} + \overline{\mathbf{c}}) + \mathbf{h}(\hat{\mathbf{x}}) - \mathbf{h}(\hat{\mathbf{x}})\| \\ &= \|\mathbf{z} - \mathbf{h}(\hat{\mathbf{x}}) + (\mathbf{a} - \mathbf{h}(\hat{\mathbf{x}} + \overline{\mathbf{c}}) + \mathbf{h}(\hat{\mathbf{x}}))\| \\ &= \|\mathbf{z} - \mathbf{h}(\hat{\mathbf{x}})\| = \|\mathbf{r}\| \leq \tau \end{aligned} \tag{6.19}$$

which concludes the proof that the attack cannot be detected by the residual-based BD detection method.

Remark: *Comparing FDIA on DC SE in the literature* [2–6] *with the FDIA on AC SE, it is observed that to successfully launch the DC-based FDIA, the hacker just needs to know the system topology, i.e.,* $\mathbf{H}$ *matrix. However, to implement the AC-based FDIA, the hacker should know both the system topology that related to the nonlinear measurement function* $\mathbf{h}(\cdot)$ *and the real-time measurements that can be used to estimate the state* $\hat{\mathbf{x}}$.

6.4.2 Practical FDIA on AC SE

Actually, due to the limited source of the hacker for obtaining real-time measurements (e.g., some measurements are protected by the control center so that the hacker does not have the power to get them), the estimated state may be different from the state the control center estimates. In other words, the state that the hacker estimates before launching the FDIA may have bias, i.e., $\tilde{\mathbf{x}} = \hat{\mathbf{x}} + \boldsymbol{\zeta}$, which is the case for practical power systems. In this scenario , the practical FDIA on AC SE should be investigated and analyzed.

Proposition 3. *If the true residual* $\|\mathbf{r}\|$ *of the uncompromised measurement can bypass the* L_2*-norm measurement residual-based BD detection and the error of the attack vector*

$$\|\mathbf{a} - \mathbf{h}(\hat{\mathbf{x}} + \mathbf{c}) + \mathbf{h}(\hat{\mathbf{x}})\| \leq \alpha = \tau - \|\mathbf{z} - \mathbf{h}(\hat{\mathbf{x}})\| = \tau - \|\mathbf{r}\| \tag{6.20}$$

holds, the malicious measurement $\mathbf{z} + \mathbf{a}$ *can also pass this detector.*

Proof *since the uncompromised measurement* $\mathbf{z}$ *can bypass the* L_2*-norm measurement residual-based BD detection,* $\|\mathbf{r}\| = \|\mathbf{z} - \mathbf{h}(\hat{\mathbf{x}})\| \leq \tau$ *holds. Thus, the* L_2*-norm residual of the attacked measurements is:*

$$\begin{aligned}\|\mathbf{r}_a\| &= \|\mathbf{z} + \mathbf{a} - \mathbf{h}(\hat{\mathbf{x}} + \mathbf{c})\| = \|\mathbf{z} + \mathbf{a} - \mathbf{h}(\hat{\mathbf{x}} + \mathbf{c}) + \mathbf{h}(\hat{\mathbf{x}}) - \mathbf{h}(\hat{\mathbf{x}})\| \\ &= \|\mathbf{z} - \mathbf{h}(\hat{\mathbf{x}}) + (\mathbf{a} - \mathbf{h}(\hat{\mathbf{x}} + \mathbf{c}) + \mathbf{h}(\hat{\mathbf{x}}))\| \\ &\leq \|\mathbf{z} - \mathbf{h}(\hat{\mathbf{x}})\| + \|\mathbf{a} - \mathbf{h}(\hat{\mathbf{x}} + \mathbf{c}) + \mathbf{h}(\hat{\mathbf{x}})\| \\ &= \|\mathbf{r}\| \leq \tau \end{aligned} \tag{6.21}$$

which concludes the proof that the attack cannot be detected by the residual-based BD detection method.

When the estimated state the hacker gets has bias before he/she starts to perform the FDIA, the attack vector changes to:

$$\mathbf{a} = \mathbf{h}(\tilde{\mathbf{x}} + \mathbf{c}) - \mathbf{h}(\tilde{\mathbf{x}}) \tag{6.22}$$

Thus, the L_2-norm residual of the attacked measurements becomes:

$$\begin{aligned}\|\mathbf{r}_a\| &= \|\mathbf{z} + \mathbf{a} - \mathbf{h}(\hat{\mathbf{x}} + \mathbf{c})\| = \|\mathbf{z} + \mathbf{a} - \mathbf{h}(\hat{\mathbf{x}} + \mathbf{c}) + \mathbf{h}(\hat{\mathbf{x}}) - \mathbf{h}(\hat{\mathbf{x}})\| \\ &= \|\mathbf{z} - \mathbf{h}(\hat{\mathbf{x}}) + (\mathbf{a} - \mathbf{h}(\hat{\mathbf{x}} + \mathbf{c}) + \mathbf{h}(\hat{\mathbf{x}}))\| \\ &\leq \|\mathbf{z} - \mathbf{h}(\hat{\mathbf{x}})\| + \|\mathbf{a} - \mathbf{h}(\hat{\mathbf{x}} + \mathbf{c}) + \mathbf{h}(\hat{\mathbf{x}})\| \\ &= \|\mathbf{r}\| + \|\mathbf{h}(\tilde{\mathbf{x}} + \mathbf{c}) - \mathbf{h}(\hat{\mathbf{x}} + \mathbf{c}) - (\mathbf{h}(\tilde{\mathbf{x}}) - \mathbf{h}(\hat{\mathbf{x}}))\|\end{aligned} \tag{6.23}$$

Perform the Taylor series expansion of $\mathbf{h}(\tilde{\mathbf{x}} + \mathbf{c})$ *and* $\mathbf{h}(\tilde{\mathbf{x}})$ *at the point* $\hat{\mathbf{x}}$*, respectively, yielding*

$$\begin{aligned}\mathbf{h}(\tilde{\mathbf{x}} + \mathbf{c}) - \mathbf{h}(\hat{\mathbf{x}} + \mathbf{c}) &= \mathbf{h}(\hat{\mathbf{x}} + \mathbf{c}) + \mathbf{J}_1(\tilde{\mathbf{x}} - \hat{\mathbf{x}} + \mathbf{c}) + o_1(\tilde{\mathbf{x}} - \hat{\mathbf{x}}) - \mathbf{h}(\hat{\mathbf{x}} + \mathbf{c}) \\ &= \mathbf{J}_1(\boldsymbol{\zeta} + \mathbf{c}) + o_1(\boldsymbol{\zeta})\end{aligned} \tag{6.24}$$

$$\begin{aligned}\mathbf{h}(\tilde{\mathbf{x}}) - \mathbf{h}(\hat{\mathbf{x}}) &= \mathbf{h}(\hat{\mathbf{x}}) + \mathbf{J}_2(\tilde{\mathbf{x}} - \hat{\mathbf{x}}) + o_2(\tilde{\mathbf{x}} - \hat{\mathbf{x}}) - \mathbf{h}(\hat{\mathbf{x}}) \\ &= \mathbf{J}_2\boldsymbol{\zeta} + o_2(\boldsymbol{\zeta})\end{aligned} \tag{6.25}$$

where $\mathbf{J}_1 = \partial\mathbf{h}/\partial\mathbf{x}|_{\mathbf{x}=\hat{\mathbf{x}}+\mathbf{c}}$*;* $\mathbf{J}_2 = \partial\mathbf{h}/\partial\mathbf{x}|_{\mathbf{x}=\hat{\mathbf{x}}}$*;* $o_1(\boldsymbol{\zeta})$ *and* $o_2(\boldsymbol{\zeta})$ *are the higher-order Taylor expansion terms, respectively. It should be noted that in the WLS-based SE algorithm, only the first-order approximation is used, which means that all the higher-order terms are neglected during the iteration. In other words,* $o_1(\boldsymbol{\zeta})$ *and* $o_2(\boldsymbol{\zeta})$ *tend to 0 after the convergence of the WLS method. Therefore,*

$$\begin{aligned}&\|\mathbf{h}(\tilde{\mathbf{x}} + \mathbf{c}) - \mathbf{h}(\hat{\mathbf{x}} + \mathbf{c}) - (\mathbf{h}(\tilde{\mathbf{x}}) - \mathbf{h}(\hat{\mathbf{x}}))\| \\ &= \|(\mathbf{J}_1 - \mathbf{J}_2)\boldsymbol{\zeta} + \mathbf{J}_1\mathbf{c} + (o_1(\boldsymbol{\zeta}) - o_2(\boldsymbol{\zeta}))\| \\ &\cong \|(\mathbf{J}_1 - \mathbf{J}_2)\boldsymbol{\zeta} + \mathbf{J}_1\mathbf{c}\| \leq \|\mathbf{J}_1 - \mathbf{J}_2\|\|\boldsymbol{\zeta}\| + \|\mathbf{J}_1\|\|\mathbf{c}\|.\end{aligned} \tag{6.26}$$

By combining (6.23) and (6.26), we get:

$$\|\mathbf{r}_a\| \leq \|\mathbf{r}\| + \|\mathbf{J}_1 - \mathbf{J}_2\|\|\boldsymbol{\zeta}\| + \|\mathbf{J}_1\|\|\mathbf{c}\|. \tag{6.27}$$

In order not to be detected by the control center, (6.27) should be less than the BD detection threshold, i.e., $\|\mathbf{r}\| + \|\mathbf{J}_1 - \mathbf{J}_2\|\|\boldsymbol{\zeta}\| \leq \tau$*, yielding the upper bound of the attack to be:*

$$0 \leq \|\mathbf{J}_1 - \mathbf{J}_2\|\|\boldsymbol{\zeta}\| + \|\mathbf{J}_1\|\|\mathbf{c}\| \leq \tau - \|\mathbf{r}\|. \tag{6.28}$$

Remark: *Equation 6.28 actually shows the tradeoff between the attack magnitude and the error of the estimated vector. Once the estimated error of the system states* *is fixed, i.e.,* $\|\boldsymbol{\zeta}\| = \beta \neq 0$*, the attack magnitude is bounded as:*

$$\begin{aligned}(\|\mathbf{J}_1\| - \|\mathbf{J}_2\|)\beta + \|\mathbf{J}_1\|\|\mathbf{c}\| \leq \|\mathbf{J}_1 - \mathbf{J}_2\|\beta + \|\mathbf{J}_1\|\|\mathbf{c}\| \leq \tau - \|\mathbf{r}\| \\ \downarrow \\ 0 \leq \|\mathbf{c}\| \leq \|\mathbf{J}_2\| - \beta + \frac{\tau - \|\mathbf{r}\| + \|\mathbf{J}_2\|\beta}{\|\mathbf{J}_1\|}.\end{aligned} \tag{6.29}$$

Here, if $\|\boldsymbol{\zeta}\| = \beta = 0$*, which means that the hacker can get exactly the same state estimate before he/she performs the FDIA, the attack reduces to the perfect FDIA on AC SE mentioned in Section* 5.4.1. *The attack magnitude is bounded according to* (6.28) *as:*

$$0 \leq \|\mathbf{c}\| \leq \frac{\tau - \|\mathbf{r}\|}{\|\mathbf{J}_1\|}. \tag{6.30}$$

This upper bound is more tight than setting $\|\boldsymbol{\zeta}\| = \beta = 0$ *directly in (6.29), i.e.,*

$$0 \leq \|\mathbf{c}\| \leq \frac{\tau - \|\mathbf{r}\|}{\|\mathbf{J}_1\|} \leq \|\mathbf{J}_2\| + \frac{\tau - \|\mathbf{r}\|}{\|\mathbf{J}_1\|}. \tag{6.31}$$

6.5 Proposed measurement consistency check-based FDIA detection method

Recall that the hacker's objective is to change the estimated state value by using malicious measurements injection strategy. In other words, if there exists FDIA measurements, the SCADA-based SE will significantly deviate from their estimated state values without attacks. That motivates to exploit alternative measurement source as back up protection so that the SCADA-based SE results could be double checked to guarantee its reliability. In Reference 23, the greedy algorithm-based secure PMUs placement strategy to defend against FDIA is developed. The main idea is to deploy a minimal number of PMUs through greedy search-based algorithm so that the construction of the attack vector is restricted. However, this method is only applicable to some specific attack scenarios, i.e., given a target number of SCADA measurements. Once the number of attacked measurements changes, the method has to be run again to determine the number of secure PMUs that should be deployed or protected. The problem is that for large-scale power systems, the greedy search-based algorithm is very time consuming and therefore it is not suitable for online application. On the other hand, when the measurements from the deployed PMUs contain BD, the performance of this method will be degraded. The same stories happen in References 24–26 that the BD in received PMU measurements would significantly affect the FDIA detection performance.

To solve these problems and make the FDIA detectable, this paper proposes a measurement consistency check-based detection method using additional deployed or protected PMU measurements. The minimal PMU placement method [27] that can ensure the observability of a given power system is used to determine the number of PMUs that need to be protected or deployed. Then, the robust Huber-estimator [28] is used to perform the PMU-based linear SE, resulting in good estimation results even when BD occurs in PMU measurements. The estimated measurements are used to check the reliability of the original SCADA measurements through a proposed statistical-based test. Note that we also assume these limited number of PMUs are secure.

6.5.1 Robust Huber-estimator for power system

In this chapter, we assume the minimal PMU placement method [27] that can ensure the observability of a given power system is adopted. All the measurements from these limited PMUs are protected so that the hacker cannot get access to them. Note that, if a given system has already deployed the enough number of PMUs that can make the system observable, we just need to protect them. Otherwise, we have to deploy new secure PMUs so that the system can be observed by the limited PMUs. As indicated in Reference 29, once the system is PMU observable, the nonlinear measurement function in (6.1) will be linear, i.e.,

$$\mathbf{z} = \mathbf{H}_p\mathbf{x} + \mathbf{v} \tag{6.32}$$

where $\mathbf{H}_p$ is a matrix that represents the linear relationship between the measurement vector $\mathbf{z}$ and the state vector $\mathbf{x}$. The elements of $\mathbf{H}_p$ are constituted by the system conductance and susceptance. The error $\mathbf{v}$ is assumed to be distributed according to the contaminated model

$$G = (1 - \xi)\mathbf{\Phi} + \xi\Delta_r \tag{6.33}$$

where $\xi \in [0\ 0.5]$; $\mathbf{\Phi}$ is the multivariate Gaussian cumulative distribution function; Δ_r is an unknown distribution function, which is not necessary Gaussian distribution. For small value of ξ, this model indicates there is a large fraction of the errors follow Gaussian distribution while the remaining fraction follows the unknown distribution. The distribution G is not Gaussian except for $\xi = 0$.

In order to obtain good results even when BD occur in PMU measurements and the error of PMU measurements does not follow Gaussian distribution, the robust Huber-estimator [28] is adopted. Using the minimax principle, Huber showed that at the Gaussian distribution, the Huber-estimator which minimizes the maximum asymptotic variance among all location equivariant estimators over the contaminated model G. The Huber-estimator is defined such that it minimizes the following objective function

$$J(\mathbf{x}) = \sum_{i=1}^{m} \rho(r_{S_i}) \tag{6.34}$$

where m is the number of measurements; $r_{S_i} = r_i/s$ is the standardized residual; $r_i = \mathbf{z}_i - \mathbf{a}_i^T\hat{\mathbf{x}}$; $\mathbf{a}_i^T$ is the ith column vector of the matrix $\mathbf{H}_p^T$; $s = 1.4826\ \text{median}_i|r_i|$ is the robust scale estimate; $\rho(\cdot)$ is the Huber-ρ function defined as:

$$\rho(r_{S_i}) = \begin{cases} \dfrac{1}{2}r_{S_i}^2 & \text{for } |r_{S_i}| < b \\ b|r_{S_i}| - b^2/2 & \text{elsewhere} \end{cases} \tag{6.35}$$

where the parameter is set as $b = 1.5$ with high efficiency at Gaussian noise [28, 30] while not increasing too much the bias under contamination.

To minimize (6.34), one takes its partial derivative and sets it equal to zero, yielding

$$\frac{\partial J(\mathbf{x})}{\partial \mathbf{x}} = \sum_{i=1}^{m} -\frac{\mathbf{a}_i}{s}\psi(r_{S_i}) = \mathbf{0} \tag{6.36}$$

where $\psi(r_{S_i}) = \partial\rho(r_{S_i})/\partial r_{S_i}$. Equation (6.36) can be arranged in the matrix form

$$\mathbf{H}_p^T\mathbf{Q}(\mathbf{z} - \mathbf{H}_p\hat{\mathbf{x}}) = \mathbf{0} \tag{6.37}$$

where $q(r_{S_i}) = \psi(r_{S_i})/r_{S_i}$ and $\mathbf{Q} = \mathrm{diag}(q(r_{S_i}))$.

Then, the iteratively reweighted least squares algorithm [28, 30] is used to solve the estimation, yielding

$$\hat{\mathbf{x}}^{(j+1)} = \left(\mathbf{H}_p^T\mathbf{Q}^{(j)}\mathbf{H}_p\right)^{-1}\mathbf{H}_p^T\mathbf{Q}^{(j)}\mathbf{z} \tag{6.38}$$

where j is the iteration counter. The iterations are stopped when:

$$\left\|\hat{\mathbf{x}}^{(j+1)} - \hat{\mathbf{x}}^{(j)}\right\|_{\infty} \le tol, \quad \text{e.g., } tol = 10^{-2} \tag{6.39}$$

Remark: *According to Huber's proposal, the final estimate of a linear regression model ($\mathbf{Y} = \mathbf{X}\boldsymbol{\theta} + \mathbf{e}$ with zero mean error distribution for example) using Huber-estimator has an approximate normal asymptotic probability distribution function with zero mean and asymptotic variance matrix* $\mathbf{V}$*, which can be derived by the influence function (IF) and is expressed as [28]:*

$$\begin{aligned}\mathbf{V} &= \lim_{m\to\infty} V\left(\sqrt{m}\hat{\boldsymbol{\theta}}_m\right) = E\left[IF \cdot IF^T\right] \\ &= \frac{E[\psi^2(r)]}{\left(E[\psi'(r)]\right)^2}\left(\mathbf{X}^T\mathbf{X}\right)^{-1}\end{aligned} \tag{6.40}$$

where m *is the number of measurement;* $E[\cdot]$ *is the expectation operator. Therefore, the asymptotic covariance matrix* $\mathbf{P}$ *of the estimation is:*

$$\mathbf{P} = \frac{E[\psi^2(r)]}{\left(E[\psi'(r)]\right)^2}\left(\mathbf{H}_p^T\mathbf{H}_p\right)^{-1} \tag{6.41}$$

where $\frac{E[\psi^2(r)]}{\left(E[\psi'(r)]\right)^2} = 1.0369$ *when the threshold for the Huber-ρ is 1.5. Thus, we can obtain* $\hat{\mathbf{x}} \sim N(\mathbf{0}, \mathbf{P})$.

6.5.2 Proposed statistical test-based FDIA detection method

When the estimated state associated with covariance matrix is obtained, the estimated SCADA measurements can be calculated

$$\hat{\mathbf{z}} = \mathbf{H}\hat{\mathbf{x}} \tag{6.42}$$

with the covariance matrix

$$\boldsymbol{\Sigma} = \text{cov}(\mathbf{H}\hat{\mathbf{x}}) = \mathbf{H}\,\text{cov}(\hat{\mathbf{x}})\mathbf{H}^T = \mathbf{HPH}^T \tag{6.43}$$

where **H** represents the relationship between the state and the received SCADA measurements.

Define the difference between the estimated SCADA measurements and the received SCADA that may be attacked as the new residual

$$v = \mathbf{z} - \hat{\mathbf{z}} \tag{6.44}$$

Proposition 4. *Assume the received PMU and SCADA measurements are from distinct devices and do not have any correlation, thus the new residual* $v = \mathbf{z} - \hat{\mathbf{z}}$ *should be normally distributed with zero mean and covariance* $\mathbf{C} = \mathbf{R} + \mathbf{HPH}^T$.

Proof Since the calculated SCADA measurements are obtained using PMU measurements-based SE results and the received PMU measurements do not have any correlation with the received SCADA measurements, the errors present in received SCADA measurements and the calculated SCADA measurements are uncorrelated. Thus, the error covariance matrix of the new residual can be calculated as:

$$\mathbf{C} = \text{cov}\left(\mathbf{z} - \hat{\mathbf{z}}\right) = \text{cov}(\mathbf{z}) + \text{cov}\left(\hat{\mathbf{z}}\right) = \mathbf{R} + \mathbf{HPH}^T. \tag{6.45}$$

On the other hand, since gross error of SCADA is usually assumed to be normally distributed with zero mean and the $E\left[\hat{\mathbf{z}}\right] = E[\mathbf{H}\hat{\mathbf{x}}] = \mathbf{H}E[\hat{\mathbf{x}}] = \mathbf{0}$, thus:

$$E\left[\mathbf{z} - \hat{\mathbf{z}}\right] = E[\mathbf{z}] - E\left[\hat{\mathbf{z}}\right] = \mathbf{0} \tag{6.46}$$

which completes the proof.

Since the residual follows the Gaussian distribution, we can define the following statistical test

$$v_{N_i} = \frac{|v_i|}{\sqrt{\mathbf{C}(i,i)}} = \frac{\left|\mathbf{z}_i - \hat{\mathbf{z}}_i\right|}{\sqrt{\mathbf{C}(i,i)}} \le \eta \tag{6.47}$$

where v_i and v_{N_i} are ith element of the residual and the normalized residual, respectively; $\mathbf{C}(i,i)$ is the ith diagonal element of the covariance matrix; η is the threshold that is determined by the confidence level. For example, $\eta = 3$ is obtained with 99.7% confidence level.

If there is no BD or FDIAs among the received SCADA measurements, (6.47) would not be violated; otherwise, BD or FDIAs exist and the corresponding measurements whose normalized residuals that are larger than the threshold could be identified as BD or attacked measurements.

6.6 Simulation results

In this section, the proposed method is tested on Institute of Electrical and Electronics Engineers (IEEE) 14-bus and 30-bus test systems. The PMU installed at a bus can measure the nodal voltage phasors as well as all the current phasors of transmission lines that are connected to this bus. In the simulation, since the PMU measurements are more accurate than the SCADA measurements, their noise is smaller compared with the SCADA measurement noise. The noise for PMU measurements is assumed to be normally distributed with zero mean and variance, 0.1% [29], while the noise for traditional SCADA measurements is assumed to be normally distributed with zero mean and variance 1%; the fast decoupled power flow program is used to obtain the system true state; the minimum energy attack residue-based attack strategy [4] is used to construct the attack vector. All the tests are performed in MATLAB® environment using Intel Core i5 2.5 Hz CPU with 8 GB memory computer.

6.6.1 Detection of FDIA on DC SE

We evaluate the proposed method using the IEEE-14 test system with minimal PMU placement, i.e., buses 2, 6, 7, and 9 are installed with PMUs, where power system observability is ensured. In the simulation, there are 13 state variables and 54 measurements. The receiver operator characteristic (ROC) curves that can characterize the tradeoff between the probabilities of attack detection and false alarm are used. The following two cases are considered:

Case 1: Two measurements for bus 11 are compromised and the objective is to introduce 10% error to the angle of bus 11.

Case 2: Three measurements for bus 5 are compromised and the objective is to introduce 10% error to the angle of bus 5.

Figures 6.1 and 6.2 show the ROC curves for different detectors, where the generalized likelihood ratio test (GLRT)-based detector, infinite norm detector,

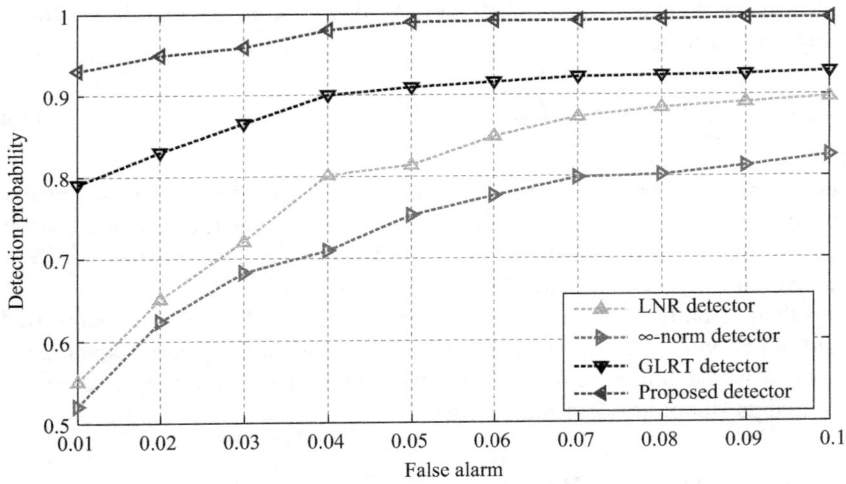

Figure 6.1 ROC curves for different detectors when two measurements are compromised in IEEE 14-bus test system

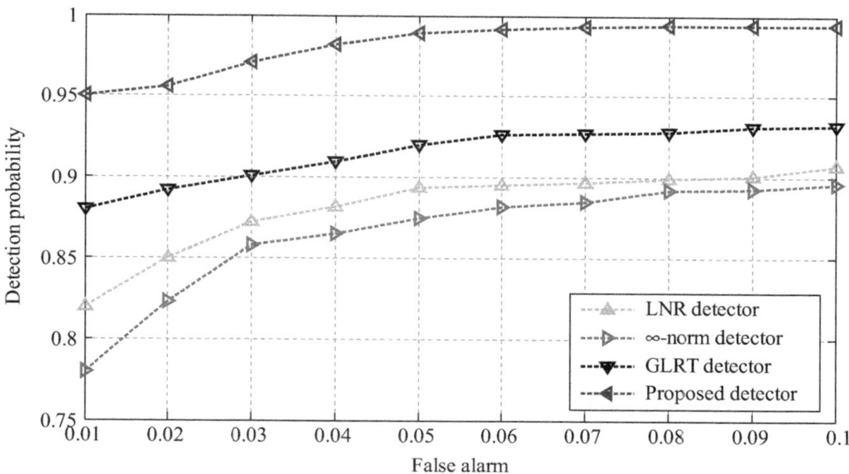

Figure 6.2 ROC curves for different detectors when three measurements are compromised in IEEE 14-bus test system

and LNR detector are presented in Reference 4. From the figures, we can find that the proposed detector can achieve a much higher detection probability than the other three detectors in both cases. For example, in case 1, for a 0.05 false alarm probability, the attack detection probabilities of the proposed detector, GLRT, infinite detector, and LNR detector are 0.982, 0.91, 0.82, and 0.75, respectively. Compared with Figures 6.1 and 6.2, it is also observed that with the increase numbers of attacked measurements, the probability to detect the attack increases. On the other hand, in both figures, even the false alarm probability is very low, e.g., 0.01, the proposed method can obtain at least 93% probability to successfully detect the attack. This means to protect 4 PMUs at node 2, 6, 7, 9 for the IEEE-14 test system could help the system to decrease the probability of being attacked by the hacker.

6.6.2 Detection of FDIA on AC SE

In this section, numerical simulations on the IEEE 30-bus test system depicted in Figure 6.3 are given to evaluate the effectiveness of proposed AC FDIA detection method. The IEEE 30-bus test system is measured by 93 SCADA measurements (18 pairs of active and reactive power injections; 28 pairs of power flows; voltage at bus 1); 10 PMUs are installed at buses 1, 7, 8, 10, 11, 12, 18, 23, 26, and 30, which is the minimal number of PMUs that can ensure the system observable; all the measurements from the PMUs are assumed to be protected by the control center so that the hacker cannot get access to them. Two scenarios are considered:

Case 3: Perfect FDIA to the phase of bus 26, where the true value is $\theta_{26} = -0.2990$ radians. The objective of the adversary is to change this value to -0.0990

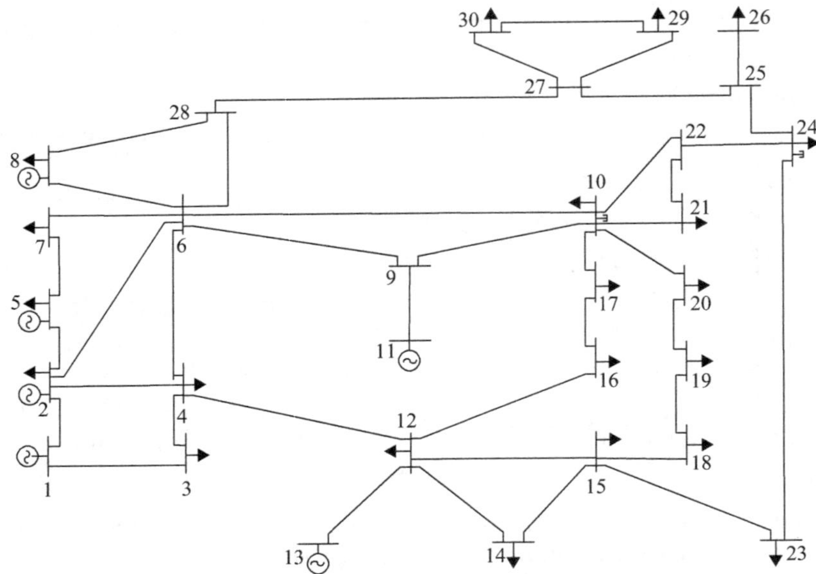

Figure 6.3 IEEE 30-bus test system

radians. To accomplish this goal, the hacker needs to compromise the measurements $P_{25-27}, Q_{25-27}, P_{25-24}, Q_{25-24}, P_{25}$, and Q_{25}.

Case 4: Imperfect FDIA to the phase on bus 26 and the object of the adversary is also to change the bus angle of 26 to -0.0990. In this case, the measurements $P_{25-27}, Q_{25-27}, P_{25-24}, Q_{25-24}, P_{25}$ and Q_{25} the hacker obtains have already had 10% error before he/she launches the attack (but the hacker does not know the error).

The ROC curves are also adopted to test the performance of different detectors, i.e., GLRT detector, infinite norm detector, LNR detector, and the proposed detector. Figures 6.4 and 6.5 show the ROC curves for different detectors.

It is observed from these two figures that for both cases, the proposed detector outperforms the other three detectors. Compared with the perfect and imperfect FDIAs, we can find that it is harder for infinite, LNR and GLRT detectors to detect perfect FDIA with high probability, while the proposed detector is slightly affected and can detect FDIA with at least 95% probability for both cases. It is obvious that when uncertainty increases in the obtained measurements by the hacker (the imperfect FDIA case), the control center is easier to detect the FDIA through traditional detection methods. Actually, the perfect FDIA can be regarded as the worst case for the detector in control center. The other three detectors do not perform well in the worst case. The proposed detector is an alternative effective method that can guarantee the reliability of the SE results with high confidence.

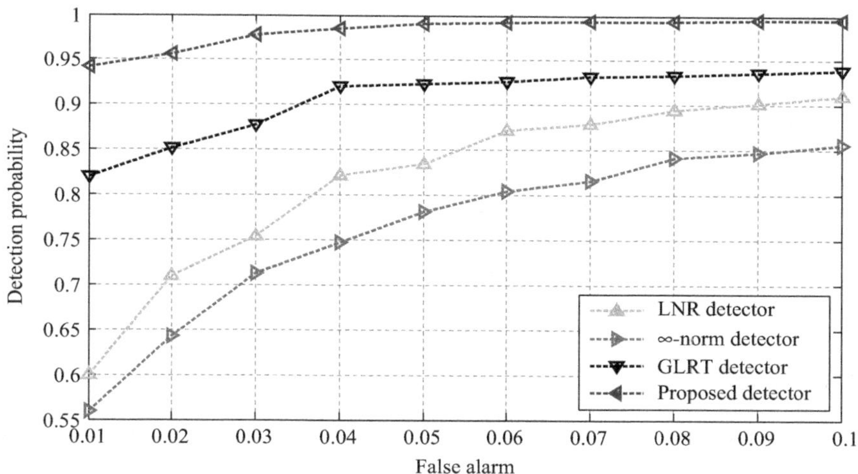

Figure 6.4 ROC curves for different detectors in IEEE 30-bus test system: Case 3 – perfect FDIA

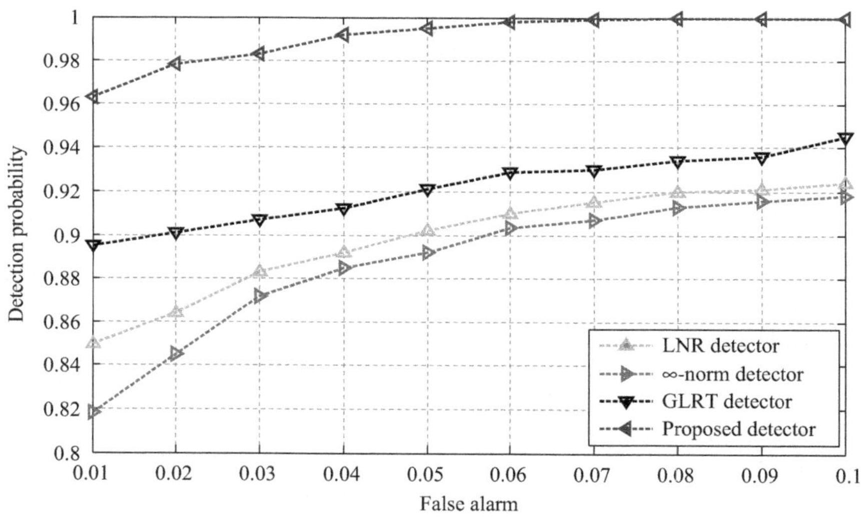

Figure 6.5 ROC curves for different detectors in IEEE 30-bus test system: Case 4 – imperfect FDIA

6.7 Conclusions

In this chapter, the authors extended the existing FDIAs on DC or AC model to a more generalized framework, where perfect and imperfect attacks are included. The vulnerability of SE under this generalized framework was investigated through the analysis of the tradeoff between attack magnitude and the attack error. To detect FDIAs, a measurement consistency check-based index is proposed. This index is defined by the measurement difference between received measurements and the interpolated measurements that are calculated through a small number of selected secure alternative PMU measurements. This index works well even when BD and FDIAs occur simultaneously. Simulation results on IEEE 14-bus and 30-bus test systems validated the effectiveness of the proposed method.

With the development of WAMS, the proposed method could be a good candidate to mitigate the effects of cyber security on power system SE and measurements-based applications.

References

[1] A. Monticelli, "Electric power system state estimation," *Proc. IEEE*, vol. 88, no. 2, pp. 262–282, 2000.

[2] Y. Liu, P. Ning, M. K. Reiter, "False data injection attacks against state estimation in electric power grids," *ACM Trans. Infor Sys Sec,* vol. 14, no. 1, pp. 1–33, 2011.

[3] S. Cui , Z. Han, S. Kar, T. T. Kim, H. V. Poor, A. Tajer, "Coordinated data-injection attack and detection in the smart grid: A detailed look at enriching detection solutions," *IEEE Signal Process. Mag.,* vol. 29, no. 5, pp. 106–115, 2012.

[4] O. Kosut, L. Jia, R. J. Thomas, L. Tong, "Malicious data attacks on the smart grid," *IEEE Trans. Smart Grid*, vol. 2, no. 4, pp. 645–658, 2011.

[5] J. Kim, L. Tong, "On topology attack of a smart grid: Undetectable attacks and countermeasures," *IEEE J. Sel. Areas Commun.*, vol. 31, no. 7, pp. 1294–1305, 2013.

[6] Y. Y. Ling, L. Z. Y. R. Kui, "Modeling load redistribution attacks in power systems," *IEEE Trans. Smart Grid,* vol. 2, no. 2, pp. 382–390, 2011.

[7] G. Hug, J. A. Giampapa, "Vulnerability assessment of AC state estimation with respect to false data injection cyber-attacks," *IEEE Trans. Smart Grid*, vol. 3, no. 3, pp. 1362–1370, 2012.

[8] M. Rahman, H. Mohsenian-Rad, "False data injection attacks against nonlinear state estimation in smart power grids," *IEEE Power Eng. Soc. General Meeting*, 21–25 July 2013, Vancouver, BC, 1–5.

[9] L. Jia, R. J. Thomas, L. Tong, "On the nonlinearity effects on malicious data attack on power system," *IEEE Power Eng. Soc. General Meeting*, 22–26 July 2013, Vancouver, BC, San Diego, CA: 1–8.

[10] J. B. Zhao, G. X. Zhang, Z. Y. Dong, K. P. Wong, "Forecasting-aided imperfect false data injection attacks against power system nonlinear state estimation", *IEEE Trans. Power Systems*, vol. 7, no. 1, pp. 6–8, 2016.

[11] H. Sandberg, A. Teixeira, K. H. Johansson, "On security indices for state estimators in power networks," *First Workshop Secure Control Systems (CPS WEEK)*, Stockholm, Sweden, April 2010.

[12] T. T. Kim, H. V. Poor, "Strategic protection against data injection attacks on power grids," *IEEE Trans. Smart Grid,* vol. 2, no. 2, pp. 326–333, 2011.

[13] L. C. Liu, M. Esmalifalak, Q. F. Ding, V. A. Emesih, H. Zhu, "Detecting false data injection attacks on power grid state estimation by sparse optimization," *IEEE Trans. Smart Grid*, vol. 5, no. 2, pp. 612–620, 2014.

[14] K. L. Morrow, E. Heine, K. M. Rogers, R. B. Bobba, T. J. Overbye. "Topology perturbation for detecting malicious data injection". *Proceedings of the 45th System Science Conference*. Hawaii, USA; 2012. p. 2104–13.

[15] M. A. Rahman, H. Mohsenian-Rad, "False data injection attacks within incomplete information against smart power grids," *Proc. IEEE GLOBECOM*, Dec. 3–7, 2012, Anaheim, CA, pp. 3153–3158.

[16] R. B. Bobba, K. M. Rogers, Q. Wang, H. Khurana, T. J. Overbye. Detecting false data injection attacks on DC state estimation. *Proc first Workshop Sec Cont Sys*. Stockholm, Sweden; 2010. p. 226–231.

[17] A. Giani, E. Bitar, M. Garcia, M. McQueen, P. Khargonekar, K. Poolla. Smart grid data integrity attacks. *IEEE Trans. Smart Grid*, vol. 4, no. 3, pp. 1244–1253, 2013.

[18] A. Abur, A. Gomez Exposito, *Power System State Estimation: Theory and Implementation*. Marcel Dekker, Inc.: New York, 2004.

[19] B. Gou, A. Abur, "An improved measurement placement algorithm for network observability," *IEEE Trans. Power Syst.*, vol. 16, no. 4, pp. 819–824, 2001.

[20] L. Mili, T. Van Cutsem, M. Ribbens-Pavella, "Bad data identification methods in power system state estimation – A comparative study," *IEEE Trans. Power App. Syst.*, vol. PAS-104, no. 11, pp. 3037–3049, 1985.

[21] L. Mili, M. G. Cheniae, N. S. Vichare, P. J. Rousseeuw, "Robust state estimation based on projection statistics [of power systems]," *IEEE Trans. Power Systems*, vol. 11, no. 2, pp. 1118–1127, 1996.

[22] Y. F. Huang, S. F. Werner, J. Huang, N. Kashyap, V. Gupta, "State estimation in electricpower grids," *IEEE Signal Proc Mag.* vol. 29, no. 5, pp. 33–43, 2012.

[23] T. T. Kim, H. V. Poor, "Strategic protection against data injection attacks on power grids," *IEEE Trans. Smart Grid*, vol. 2, no. 2, pp. 326–333, 2011.

[24] K. C. Sou, H. Sandberg, K. H. Johansson, "Data attack isolation in power networks using secure voltage magnitude measurements," *IEEE Trans. Smart Grid*, vol. 5, no. 1, pp. 14–28, 2014.

[25] J. Kim, L. Tong, "On phasor measurement unit placement against state and topology attacks," *IEEE International Conference on Smart Grid Communications*, 2013, 396–401.

[26] M. Gol, A. Abur, "Effective measurement design for cyber security," *IEEE Pow Sys Comput Conf*, 18–22 August 2014, Wroclaw, pp. 1–8.

[27] T. L. Baldwin, L. Mili, Jr. M. B Boisen, R. Adapa, "Power system observability with minimal phasor measurement placement," *IEEE Trans. Power Systems*, vol. 8, no.2, pp.707–715, 1993.

[28] F. R. Hampel, E. M. Ronchetti, P. J. Rousseeuw, W. A. Stahel, *Robust Statistics: The Approach Based on Influence Functions*. New York: John Wiley & Sons, Inc., 1986.

[29] A. S. Costa, A. Albuquerque, D. Bez, "An estimation fusion method for including phasor measurements into power system real-time modeling," *IEEE Trans. Power Systems*, vol. 28, no. 2, pp. 1910–1920, 2013.

[30] M. Gandhi, L. Mili, "Robust Kalman filter based on a generalized maximum-likelihood-type estimator, *IEEE Trans. Signal Processing*, vol. 58, no. 5, pp. 2509–2520, 2010.

[31] IEEE Std. C37.118.2011–1, *IEEE Standard for Synchrophasor Measurements for Power Systems*, revision of the IEEE Std. C37.118.2005, December 2011.

Index